MW00844491

# ARMCHAIR PHYSICS

ARMCHAIR
GUIDES

# ARMCHAIR PHYSICS

EVERYTHING YOU NEED TO KNOW,
FROM ENTROPY TO GRAVITY

ISAAC McPHEE

Brimming with creative inspiration, how-to projects, and useful information to enrich your everyday life, Quarto Knows is a favorite destination for those pursuing their interests and passions. Visit our site and dig deeper with our books into your area of interest: Quarto Creates, Quarto Cooks, Quarto Homes, Quarto Lives, Quarto Drives, Quarto Explores, Quarto Gifts, or Quarto Kids.



This edition published in 2018 by Chartwell Books,
an imprint of The Quarto Group,
142 West 36th Street, 4th Floor,
New York, NY 10018, USA
**T** (212) 779-4972 **F** (212) 779-6058
**www.QuartoKnows.com**

Conceived, designed and produced by Quid Publishing,
an imprint of The Quarto Group,
The Old Brewery,
6 Blundell Street,
London N7 9BH
United Kingdom



Chartwell Books titles are also available at discount for retail, wholesale, promotional, and bulk purchase. For details, contact the Special Sales Manager by email at specialsales@quarto.com or by mail at The Quarto Group, Attn: Special Sales Manager, 401 Second Avenue North, Suite 310, Minneapolis, MN 55401, USA.

10 9 8 7 6 5 4 3 2 1

ISBN: 978-0-7858-3597-4

Printed in China

# CONTENTS

# WHAT IS PHYSICS?

"Physics" is derived from the Greek word *physis*, which means "nature." Indeed, this is what physics is all about—the quest to explore the mysteries of how nature behaves. The purpose of physics is to ask every question that can be asked about how absolutely everything works, and then to search for the answers to these questions. In truth, there seem to be no clear limits to the scope of physics: it seeks to explain energy and matter, and how these things work together; it investigates natural phenomena both massive (such as the entire universe) and unimaginably small (such as the tiniest pieces of the atom). The study of physics is extensive and carries with it a wealth of fascinating mysteries, which is what makes it so exciting.

## FROM QUARKS TO QUASARS

Few fields of study are larger in scale than physics. Within this all-encompassing science lie answers to questions about the nature of the universe itself: its shape, its content, and its history. By applying the same physical principles by which we look at the universe as a whole, we can to gaze into the mysteries of the tiniest specks of matter. With the same enthusiasm a physicist can observe the light from distant galaxies known as quasars, the most luminous objects in the known universe, as well as the interaction between quarks, the almost infinitesimally small particles within the atomic nucleus.

Physics lies at the root of all other sciences. Indeed, everything in the universe, whether or not it can be seen, can be reduced down to the most basic physical laws. Every field of study that concerns itself with the material world—from chemistry and biology to astronomy and even engineering—is ultimately nothing more than physics. Chemistry is the study of the chemical elements (atoms), their properties, and how they bond together to form compounds and substances. Biology is the study of living creatures, made up of cells, which are made up of atoms. Engineering is the study of materials, strengths, and forces. All three of these fall within the study of physics.

"To those who do not know mathematics it is difficult to get across a real feeling as to the beauty, the deepest beauty, of nature." —*Richard Feynman*

## HOW PHYSICS IS DONE

With such a grand scope, how can physicists even begin to answer all of the questions posed by the universe? The answer lies in the fact that within physics there are multiple very specific disciplines, and most physicists today have taken up very specialized approaches to their chosen field. These are just a few of the largest disciplines:

**Particle physics** looks at the smallest things in the universe: atoms and subatomic particles. It is one of the most exciting scientific fields today; particle physicists use both theory and experiment to explore the mysteries hidden within the tiniest pieces of matter in the universe, seeking to answer questions about the origins of the universe and the basic building blocks of all matter.

**Astrophysics** (also known as "high-energy physics") explores the universe, looking at phenomena such as stars, galaxies, black holes, quasars, pulsars, and supernovas. It applies principles such as relativity and quantum mechanics to objects in our solar system and beyond.

**Nuclear physics** uncovers the potential hidden within the atomic nucleus, specifically for use in developing newer and more efficient forms of energy. From nuclear physics have come nuclear weapons and nuclear power and it has led to numerous advances in medicine, engineering, and even exploring our own history (using radioactive dating techniques).

Within each of these fields physicists can generally be divided into two broad categories: Theorists and experimenters. Theorists seek to utilize the incredible power of very complicated mathematical tools in order to get to the bottom of physical mysteries. They use tried and true methods of calculation in order to take previous experiments and observations and to generalize them, predicting future developments or simply better explaining those phenomena which have already been observed.

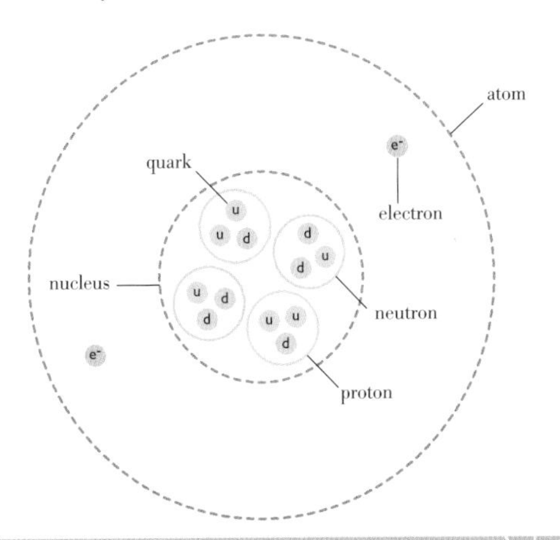


• The study of physics at a subatomic level has revealed atoms to be made up of even smaller particles, among which are six types of quarks (see p. 161).

Experimenters seek to explore the physical universe by putting it to the test. Using equipment as simple as a microscope or as complicated as a multi-billion-dollar particle accelerator, experimenters must be clever, creative, and, above all, precise. They are able to manipulate and observe the smallest bits of matter known to man and observe the farthest reaches of space with unbelievable accuracy, giving us all a more detailed and accurate picture of our universe. Experimental physicists throughout modern history have been on the cutting edge of technology, using the mental and technological tools at our disposal to provide new insights into the workings of nature.

• Modern physics has thrown up plenty of paradoxical and counter-intuitive concepts such as antimatter and black holes (see p. 165).

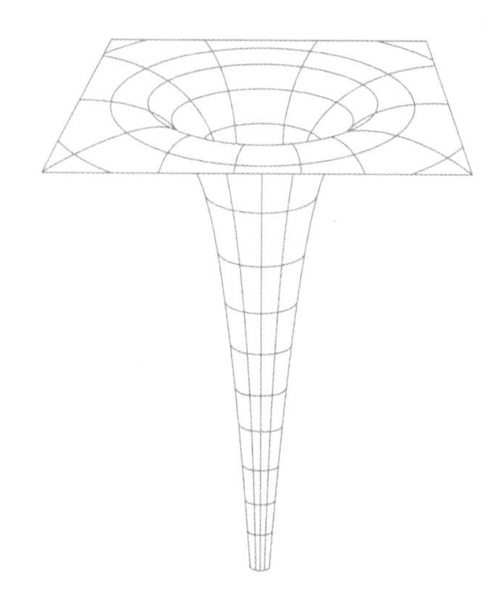

## ASKING THE RIGHT QUESTIONS

Although physicists address themselves to the most perplexing concepts in the natural world, it is important that they first learn which questions to ask. Asking the right questions can lead them to expose even greater mysteries, which is where discovery begins. Physicists throughout history have begun by asking the most basic questions and then proceeded to delve deeper into the problems that are uncovered.

We can ask, as Isaac Newton did, what makes an object fall to Earth, which has its answer, of course, in the force of gravity. But this leads us to further questions: What is gravity? Why do objects attract one another? How does the force of gravity relate to the other forces we experience, such as magnetism?

As the physicist attempts to answer these questions about the mysterious force of gravity they are led down a path that results in further questions of even greater difficulty—and there does not yet appear to be any bottom to this particular rabbit hole. Fortunately, most physicists today are not particularly concerned with finding an answer to every question about nature; they are far more interested in throwing themselves into these problems and seeing just how deep into the hole they can go. The excitement of physics, like any great adventure, comes from unveiling ever greater mysteries, and in physics, there seems to be no end in sight!

## A BRIEF HISTORY OF PHYSICS

### The Renaissance

While this book begins by exploring some of the ancient physicists who performed some very important work prior to the Middle Ages in Greece and the Middle East, in truth, the study of physics did not truly begin until the Renaissance period in Europe around the sixteenth century.

The astronomical work of Nicolaus Copernicus, Johannes Kepler, and Galileo Galilei paved the way toward a new way of looking at science itself, whereby observation, experiment, and scientific reasoning superseded previously held biases and beliefs.

With a fairly accurate picture of the true nature of our solar system firmly in place and widely understood by the seventeenth century, the scientific community in Europe began to grow and thrive, culminating in the inestimable work of Isaac Newton in the latter part of that century and into the next. Newton changed the world of science forever, taking Galileo's work in mechanics (the science of moving things) and codifying it, discovering the law of gravitation, inventing calculus (the form of mathematics which remains the backbone of theoretical physics to this day), and inspiring other scientists to believe that nature could be tamed. It could be calculated, studied, and put to use.

The work of Newton would drive physics for more than 200 years (and in many senses continues to play a vital role in science today). Throughout the eighteenth and nineteenth centuries, Newton's work was the most important ever performed in physics. Even as men like Michael Faraday and James Clerk Maxwell began to finally tame the phenomenon of electrical force, they did so within the confines of the physical laws Newton had set forth.

The universe prior to the twentieth century was a Newtonian one, a massive mechanical machine (a "clockwork universe," as it is often called) which operates under a very precise and predictable set of rules. In the Newtonian universe we had the potential to know and to predict everything. Understand the rules well enough and nothing would remain a mystery, as everything—whether animate or inanimate, man or machine—worked like a giant clock, wound up at the beginning of time and set in motion, whether by God or by nature.

"Science cannot solve the ultimate mystery of nature. And that is because, in the last analysis, we ourselves are part of nature and therefore part of the mystery that we are trying to solve." —*Max Planck*

## The Revolutions of the Twentieth Century

The Newtonian universe was shattered beyond all recognition in the twentieth century, primarily as a result of the work of Albert Einstein. While still young and practically unknown, Einstein introduced the world to the theory of relativity in 1905. Within this theory, the ideals of Newton and his followers began to show the first signs of cracks—certain things which had always been taken for granted as being absolute, such as space and time, were dramatically called into question. According to Einstein, both space and time were malleable and changed depending on one's perspective. Time could speed up and slow down, and an object's length could expand and contract along with the very fabric of space-time!

But Einstein wasn't content merely in forcing a re-examination of space-time. A decade later he would even rewrite Newton's theory of gravitation, developing an even more perfect theory in which gravity was caused by the continual "warping" of the very fabric of space-time. The revolution had begun!

The second stage of this unintentionally anti-Newtonian crusade came in the form of "quantum physics." Based on a relatively simple theory developed by the German scientist Max Planck in 1900, further refined by Einstein, Niels Bohr, and others during the 1900s and 1910s, this entirely new system of physics reached the peak of its early success in the 1920s.

In quantum physics, the cracks forming in Newton's idea of a "clockwork" universe turned into canyons, shattering the notion entirely. These new physicists realized, to the world's utter surprise, that exploring matter more deeply did not lead to more precision, as Newton would have suggested, but less. Where particles exist, nothing is determined—a notion embodied in Werner Heisenberg's famous "uncertainty principle" (see p. 138).

The rise of quantum physics led to entirely new ways of looking at physics

• The first half of the twentieth century was a period of unparalleled scientific advance, with the likes of Albert Einstein, Niels Bohr, and Werner Heisenberg leading the charge.

"Look deep into nature, and then you will understand everything better."

***—Albert Einstein***

and to new methods of performing calculations and making predictions. These new methods led to the discovery of some very peculiar things: the discovery of antimatter; the realization that even some of the smallest particles within atoms are actually made up of even smaller particles; the discovery that our universe is filled with a mysterious substance known as "dark matter"; and much, much more.

Perhaps most important, however, is the revelation that, at the smallest levels, the universe is not determined. The behaviors of particles are not predictable, as their movements are based on nothing more than probability. While this understanding has led many to speculate on some very peculiar philosophical implications, likening quantum physics to eastern religions and platonic philosophy, in the end it is just one more reminder that no matter how much we think we may know, there will always be mysteries yet to uncover in the universe. No wonder then that physics is so exciting!

## GREAT ACHIEVEMENTS IN MODERN PHYSICS

**1514**—Nicolaus Copernicus begins work on a "Sun-centered" model of the universe.

**1632**—Galileo publishes "Dialogue Concerning Two Chief World Systems," popularizing the Copernican Theory.

**1687**—Isaac Newton publishes *Philosophiæ Naturalis Principia Mathematica*, providing the world with his laws of motion, gravitation, and a fully functional theory of physics.

**1802**—John Dalton discovers the atom.

**1861**—James Clerk Maxwell provides a mathematical description of light.

**1896**—Henri Becquerel discovers radioactivity.

**1900**—Max Planck "invents" quantum physics.

**1905**—Albert Einstein's "miracle year": discovery of theory of relativity and refinement of quantum physics.

**1913**—Niels Bohr applies quantum physics to the atomic model and founds modern quantum mechanics.

**1927**—Heisenberg's uncertainty principle is unveiled.

Chapter

# Ancient Physics

Although they did not use the term "scientists," the thinkers of Ancient Greece in its golden age, and those of the Middle East and beyond, proved themselves to be thinkers of the highest caliber. This chapter offers a brief guide to those brilliant thinkers who endeavored to answer many of the same questions which continue to drive the study of physics today—questions of substance, motion, and our place in the universe.

Profile

# Thales of Miletus

**A search for the first ever scientist (although until a couple of centuries ago they were known merely as "natural philosophers") can surely lead to none other than Thales (ca. 624–546 BC). This renowned wise man was one of the "Seven Sages of Greece" and the founder of the very first known philosophical "school." Located on the coastline of modern-day Turkey in the Ionian city of Miletus (from which it took its name), the Milesian school and its students became the cornerstone of early scientific thought.**

## The First Theories of Matter

From Thales and his fellow philosophers—most notably from two of his students, Anaximander (ca. 610–546 BC) and Anaximenes (ca. 585–525 BC)—come the first theories of the material universe. The essence of the matter-theory, which stemmed from the thoughts of these contemplative men of old, is essentially the search for the "quintessential material substance"—what the Greeks called the *arche*. By today's standards, the search was not a practical one. They were not searching for the truth of matter so that they could better

### Change Is Illusory: The Philosophers Who Opposed Thales

- **Parmenides** *(early fifth century BC): The founder of the Eleatic School, Parmenides did not believe in the "change" espoused by Thales. He believed in precisely the opposite–that change is impossible, illogical, and illusory. Parmenides' beliefs heavily influenced later philosophers such as Democritus, Plato, and his immediate successor, Zeno of Elea.*

- **Zeno of Elea** *(mid-fifth century BC): Zeno is surely most remembered today for having developed a number of logical paradoxes which seek to prove, by way of a debating technique called* reductio ad absurdum *(reduction to absurdity), that change is impossible. It was not until the development of modern mathematical tools of calculus in the seventeenth century that Zeno's paradoxes could finally be given full answers.*

- **Melissus of Samos** *(fifth century BC): The third member of the Eleatic school, Melissus of Samos stressed, perhaps even more emphatically than his peers, the importance of believing in the physical world as constant, rather than changing, referring to all of what is as "The One." He argued that reality was infinite and could not change or deviate.*

"Philosophy begins with Thales."

***—Bertrand Russell***

understand subjects like chemistry or biology for pragmatic use, but instead took a more philosophical approach.

Thales considered that the essential substance was water, for in his observations water played an essential role in the formation of all other types of matter. To Anaximander the essential substance was a theoretical substance which he called *apeiron*, and to Anaximenes it was air. At the heart of the Milesian theory was the idea of change: If all could be reduced to a single substance, then there had to be a constant and unstoppable change taking place within matter in order to transform one element into another. With the Milesians, science and philosophy first began intermingling.

## Further Contributions

Though the details of his life are cloudy at best, it can be said with some certainty that Thales was the first in a long line of prominent Greek thinkers. In addition to his theories of matter (which played an essential role in spurring on the great debate that quickly followed on this subject), Thales was the first to encourage the Greek world to seek out natural explanations for phenomena rather than seeking answers only from divinity or mysticism. At the same time he made great strides in mathematics, particularly in practical geometry, three centuries before Euclid would develop the first full system of geometrical proofs. As one of the earliest astronomers, it is said that Thales famously predicted the solar eclipse that brought to a halt a battle between the Lydians and the Medes (see "Dating the Work of Thales" above). In addition to all of this was Thales' role as one of the Seven Sages of Greece. This group served as a wise and respected council to the people of the region surrounding Miletus and dispersed wisdom (though exactly how this may have looked is not understood particularly well).

### DATING THE WORK OF THALES

During which years did Thales perform his work? Those early biographers who spoke of him unfortunately provide us with very few clues as to just when all of these achievements occurred. Fortunately, historians and astronomers have been given at least one key clue: Thales' prediction of the eclipse which marked the end of a battle between the Lydians and the Medes. Turning back the clock, scholars have discovered that the eclipse in question might have been that which occurred on May 28, 585 BC, giving us at least one concrete piece of evidence for dating the work of Thales.

# THE FIRST THEORIES OF MATTER

One question dominated debate between Greek philosophers and scientists in ancient times: What is matter? These thinkers desired to understand physical reality at its most basic level, asking questions about the very substance from which everything was made. Were there many elements which combined to form matter or was it all just one? Was matter made of atoms or fluids? The questions asked by these philosophers continue to be some of the most basic problems in physics today.

**The Atomists**

Democritus performed most of his work somewhere between 440 and 400 BC. Though he is certainly best remembered today for his remarkably prescient first theory of "atoms" (which comes from the Greek word for "uncuttable"), he was known by his contemporaries and successors as the "laughing philosopher" for the cheerfulness and optimism espoused within his philosophy (the vast majority of which is lost).

Democritus used the image of a sandy beach to explain his theory of atoms. Just as the beach when viewed from afar looks like a single substance, so also might all matter be made up of tiny little granules of matter. The smallest of these pieces of matter he called "atoms."

Though today we know that there is some truth to the theory of Democritus, his work was not so easy to accept at the time, especially when so many other, apparently more logical theories existed. This first atomic theory would be neglected for more than 2,000 years before being picked up again by the likes of Galileo Galilei and Isaac Newton, and it was not fully accepted until well into the nineteenth century!

**The Elementalists**

More satisfying to the Greek world than the atomic theory of Democritus was the theory of Empedocles. Hailing from Agrigentum on the island of Sicily, Empedocles' best known achievement was a theory of matter based on a system of just four primary elements: earth, fire, air, and water. While this in itself was certainly not a revolutionary assertion (the Milesians had also considered the importance of these four elements), Empedocles was able to codify this model and turn it into a nearly complete (or so it seemed) theory of everything, whether it be matter or force.

Empedocles found himself in agreement with some of his philosophical predecessors, such as Parminides and Zeno on several counts. For example, his theory acknowledged the impossibility of existence rising from non-existence (something cannot come from nothing). However, Empedocles viewed change not as the rearrangement of tiny atoms, but as the mixture and separation of the four fundamental elements. These processes were driven by two particular forces, which he called love and strife. Appropriately enough, it was the force of love by which elements were bound together, while strife tore them apart. How poetic!

## Plato's Solids

While it was Empedocles who first began to piece together this highly rational elemental model and all of its conclusions, it was the legendary Plato—that most famous student of Socrates, born sometime around 428 BC (not long after Empedocles' death)—who advanced this theory by introducing the idea of "Platonic Solids."

Ever the master of deductive argumentation (beginning from a broad point and logically narrowing his thoughts in order to come to a specific conclusion), Plato approached matter-theory by attempting to explain Empedocles' theory through the mathematics of geometry.

Plato looked at the four elements of Empedocles and determined that each of these might be represented by a geometrical figure. These shapes—the tetrahedron (pyramid with a triangular base), hexahedron (cube), octahedron (8-sided regular solid), dodecahedron (12-sides), and icosahedron (20-sides)—were peculiar among geometrical figures in that they were the only known shapes which could be constructed whose faces, sides, and angles were all identical to one another.

"Nothing exists except atoms and empty space; everything else is opinion."

***—Democritus***

Plato argued that the four elements were not continuous substances, as Empedocles had implied, but were, as in atomic theory, made of microscopically tiny particles—in this case geometrical shapes—and that the sizes and peculiar shapes of these particles were the only things that allowed the vast differences in the elements.

### THE SCHOOLS OF GREEK PHILOSOPHY

**Milesian School**: Thales, Anaxagoras, Anaximander

**Eleatic School** (Opposed the Milesians): Parminides, Zeno, Melissus

**Atomists** (Influenced by the Eleatics): Leucippus, Democritus

**Elementalists**: Empedocles, Plato (combined elementalism and atomism), Aristotle, and many others

**Profile** # Aristotle

**Possibly the most important philosopher in all of antiquity, Aristotle's vast catalog of works covers a prodigious range of subjects, from biology and physics to medicine and theology. While many of his works were filled with brilliant insights and original research (especially in areas such as biology and moral philosophy), it was in his unique ability to outline and clarify science as a whole that Aristotle surely had the greatest impact on the course of physics in the western world.**

## The Remarkable Life of Aristotle

Aristotle was born and raised in the town of Stageira, in the eastern Greek province of Chalcidice around 384 BC. At 18 years old, Aristotle joined Plato's famed academy of thinkers in Athens, where he remained for nearly two decades until the death of his teacher. Because of his time there, much of Aristotle's thinking on philosophy closely resembles that of Plato, although he certainly went far beyond his teacher in many subjects.

Leaving the Academy and the city-state of Athens behind, Aristotle traveled to Asia Minor, where he studied two of his most beloved subjects, botany and zoology. By this point he had clearly already begun to make a name for himself in the world of philosophy because he was soon enough invited personally by King Philip II of

**Key works**

• **Organon (Logic):** *A collection of six fundamental works on logic which constitutes a full, structured system of reason.*

• **Physics:** *Aristotle's broad, far-reaching attempt at explaining every natural phenomenon in a single work; an application of his methods of logic applied to the material world.*

• **Metaphysics:** *One of Aristotle's most important works (and one of the most important philosophical works in history),* Metaphysics *goes a step further than* Physics *by asking deeper questions of the nature of existence, cause, and change. Aristotle argues that all physical objects are made up of two elements: form and matter.*

• **Ethics and Politics:** *Like many of the philosophers who followed him, many of Aristotle's purely philosophical works focused on questions of ethics and morality. In his work on ethics, Aristotle lays out a system in which a person can become virtuous only as a result of their actions. He placed ethics as a practical, rather than theoretical, pursuit.*

Macedon to become the personal tutor to his son, a young man named Alexander, who would become the famous Greek king, Alexander the Great.

Aristotle's new student would not become a force in the philosophical world himself, as perhaps the teacher had hoped, but he certainly would have considerable impact upon the whole of western civilization when, after the death of his father, he took the throne and proceeded to conquer much of the known world. So great was the empire of Aristotle's prized student that his eventual death would cause great chaos and social upheaval, extending throughout most of the Mediterranean world and beyond. It is at the death of Alexander where many historians date the end of the "classical" period of Greek history and with it much of the great philosophy of the fifth and fourth centuries.

## Aristotle's Physics

Aristotle's works (specifically his books *Physics* and *Metaphysics*) covered a vast number of subjects, reiterating and critiquing the theories of past philosophers while at the same time developing entirely new (and quite revolutionary) ideas from the ground up.

Aristotle stood by Empedocles in his elemental view of matter, though he saw fit to add a fifth element (quintessence) to the list, to which he gave the name "aether" (see *The Aether* right).

Aristotle's vast contributions to science also explored such things as the origins of the universe (where he developed a four-fold system of causes), light, optical phenomena, mathematics, biology (he developed one of the first systems for classifying plants and animals), medicine, and much more.

### THE AETHER

Aristotle's most notable addition to the theory of matter was his insistence upon the existence of the "aether," a universal, all-encompassing substance in which all other elements exist (and which holds the heavenly bodies in the sky). Though there was no particular evidence for the existence of the aether, this idea was resilient enough to survive even into the twentieth century as a substance which enabled the movement of light through space called "luminiferous aether." Finally, in 1905, Albert Einstein finally provided the world with a theory of light propagation which did not require the existence of the aether. Aristotle's grand idea was finally laid to rest after well over 2,000 years.

So venerated was Aristotle, both by his contemporaries and successors, that his contributions to western philosophy (even those contributions which were entirely incorrect) would remain mostly unchallenged for nearly 2,000 years—to question Aristotle was something very close to heresy.

**Profile**

# Archimedes

**While many of his contemporaries in ancient Greece preferred questions of abstract theory, Archimedes was a scientist of the most practical nature. He chose not to join his contemporaries in considering the nature of matter and existence; instead he attempted to better understand the world through mathematics and via the principles of simple machines—devices intended to take advantage of physical laws in order to make work easier. These machines had been in use from the earliest days of human civilization and Archimedes sought to improve upon them by applying newly developed scientific and mathematical principles.**

## The Pragmatic Philosopher

Archimedes was born in Syracuse, on the island of Sicily, sometime in the early third century BC (probably around 287) and lived for about 75 years, though much of his life remains a mystery. His life began more than a century after the period of Aristotle, during a time in which Greece existed as a number of independent city-states, Syracuse being one of them. Archimedes was therefore a citizen of Magna Grecia (the Greek-controlled regions of what is today southern Italy).

Archimedes' father, it is presumed (based on the scant records that have survived), was an astronomer named Phidias. Perhaps it is from him that Archimedes received his unceasing

**Key works**

- **On the Equilibrium of Planes:** *The work in which Archimedes explains his theories of levers.*
- **On the Measurement of a Circle:** *The work in which is explained Archimedes' new calculation of pi.*
- **On Floating Bodies:** *In this important two-volume work, Archimedes discusses the law of the equilibrium of fluids and offers the first public explanation of the principle of buoyancy. He also describes how water will form a sphere around a center of gravity.*
- **The Sand Reckoner:** *A fascinating work in which Archimedes outlines his views on astronomy and also proposes to calculate the number of grains of sand that would fit inside the known universe. The answer to this puzzle? Why, 8 x $10^{63}$ of course.*

interest in nearly every scientific and mathematical subject. Of course, Archimedes' brilliance, ingenuity, and ability to aid the Syracuse military with his many inventions did not, in the end, save his own life. He was killed by Roman soldiers during the invasion of Syracuse in 212 BC, supposedly after refusing to heed the shouts of the invading army as he studied some sort of mathematical diagram he had drawn upon the dust of his floor (his famous last words: "Do not disturb my circles").

## The Science of Archimedes

Though a great many of Archimedes' greatest and most lasting accomplishments lie in the field of mathematics (he is known as the original father of integral calculus and one of the first people to calculate the value of pi), he was also a great scientist, perhaps best remembered for his discovery of the principle of buoyancy. The story is well known.

A new golden crown had been made for King Hiero II, but there was reason to suspect that the crown might have been made of silver, rather than pure gold. Archimedes was given the task of determining the composition of the crown without destroying the object. The crux of the problem was that the density of the crown could not be determined without precise measurements of both its volume and its weight, yet because the crown's shape was so irregular, no exact calculation of its volume could be made. As the famous anecdote goes, Archimedes was perplexed by this task until, while taking a bath, he recognized that he could calculate the volume of his own irregularly-shaped body by calculating the displacement of the water in the bathtub. Upon realizing that he could use this same method to calculate the volume of the crown, the legend goes that Archimedes leaped from the bath and ran naked through the streets shouting *eureka*! ("I have found it").

## The Inventions of Archimedes

As a practical scientist of the first order, Archimedes is also well known for his many ingenious inventions, which include:

**Archimedes' Screw**— A simple device consisting of a screw-shaped blade within a hollow cylinder. When the screw is rotated, matter, such as liquid, can be carried from one end of the cylinder to the other. It is said Archimedes invented the screw to remove the bilge water from the bottom of Greek vessels.

**The Archimedes Heat Ray**—While it is very unlikely that such a device could have been constructed by the people of Syracuse, modern research has shown that the theory behind the heat ray is built upon a firm scientific foundation.

**Various Military Weapons**—It is said that Archimedes greatly improved upon the ancient idea of the catapult and designed several other key siege weapons used in the ancient world.

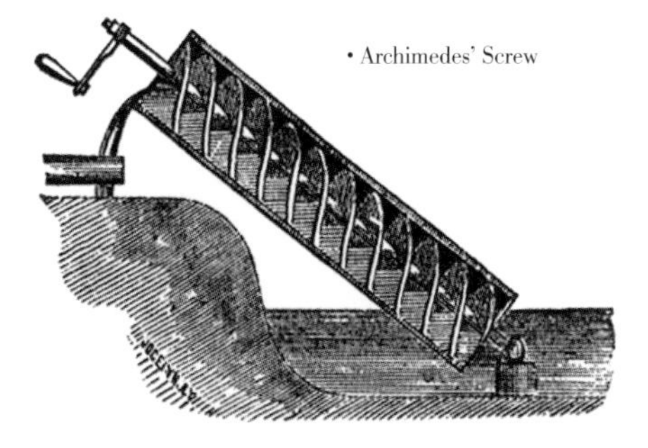

• Archimedes' Screw

# Work, Mass, Power, and Energy

## THE PROBLEM:

You are standing in front of a large, unfinished pyramid. At your feet is the one, massive brick which is needed in order to complete this impressive structure. Leading up to the top of the pyramid (where this brick must end up) is a long wooden ramp. In this example, can you isolate the following terms: work, mass, energy, and power? Can you determine how they all relate to one another?

## THE METHOD:

In the ancient world, these terms constituted the "practical" side of physics. Rather than asking broad, existential questions about matter and origins, the key questions for practical ancient thinkers were those which made work easier. With this approach in mind, the key elements of ancient physics really can be reduced to just four key concepts:

- **Work** is an action. It is using energy to accomplish a certain task (such as moving an object from one place to another).
- **Mass** refers to an object's resistance to being moved. An object of a large mass is much harder to move than an object of little mass (on the other hand, an object of a large mass is also often harder to stop moving after it has started, such as in the case of a car rolling down an icy hill—this is called momentum).
- **Energy** is what is required to do work. It is like the currency which must be paid in order for a task to be accomplished, and that which ancient workers did all they could to maximize. Energy is somewhat complex, in that it can exist in a variety of forms, such as

- mechanical (work done by machines or the human body), chemical (work by way of a chemical reaction), heat (such as in a steam engine), or even sound.
- **Power** is how quickly work can be done; in other words, if $x$ amount of energy is applied to a certain task, $y$ amount of work can be accomplished per second (or minute, or hour).

## THE SOLUTION:

The solution to this problem is relatively straightforward on the surface, although on closer inspection it does become a little more complex.

The work in this case is simple: it is the task which is to be undertaken. It will be the utilization of energy to move a large brick from one point to another.

Next, the mass. In this example, the mass relates to the massive brick which must be moved to the top of the pyramid. This is the crux of your problem and the reason you're so anxious about this particular task.

Energy is the "currency" required in order to move this pyramid. In this case, the form of energy will most likely be mechanical (using the machine of the human body) unless you are clever enough to develop a pneumatic (physical) or steam-powered (heat) device to aid you in your task. Though the form of energy may be relatively simple to point out, it is the amount of energy needed to accomplish any task which can be complicated to calculate, for ancient workers surely learned very quickly that the amount of energy required for a task depends not only on the mass of the object, but on the surface over which it is being slid (less friction = less energy required), the steepness of the surface (less steep = less energy required), the direction of the applied force, and any number of other factors.

Finally we have power. Power refers directly to you, the mover of the brick. How much energy can you exert upon that brick per second in order to slide it along the ramp? How much power you are able to exert upon the brick is a measure of just how efficiently you will be able to do the work and accomplish the task. Power is a measure, ultimately, of how successful you will be.

Perhaps neither the Egyptians, nor the Incas, nor the builders of Stonehenge truly thought of the tasks in terms such as these, and yet they learned to complete them through trial and error, experimentation and reasoning—all elements of what even today we would call the "scientific method."

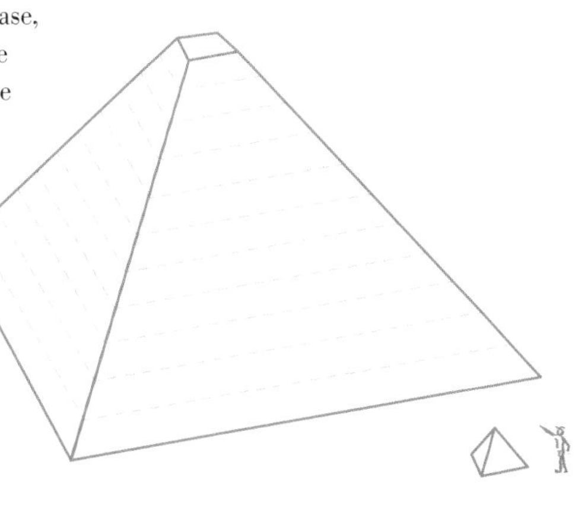

# Exercise 2 Moving the World with a Lever

**THE PROBLEM:**

Archimedes made a famous declaration concerning the lever: "Give me a place to stand and I shall move the world." By this he simply meant that with a large enough lever (and a place to stand in outer space), a single person should be able to move the mass of the entire world! But how far from Earth would someone have to be in order to make this feat theoretically possible?

Imagine that Archimedes himself, weighing 150 pounds (68 kg), was able to fly into space with his giant lever. With the fulcrum of his lever placed exactly 10,000 miles (16,000 km) from the Earth, what is the total length of the lever needed by Archimedes in order to move the world?

• Archimedes imagined himself somewhere in the farthest reaches of space, using a lever to move the Earth to demonstrate the power of simple machines.

## THE METHOD:

To even begin to solve this problem Archimedes will need to cover a few basics. First, how does he calculate the abilities of a lever? Fortunately, there is a rather simple formula used for this:

$$L1 = \frac{(m2 \times L2)}{m1}$$

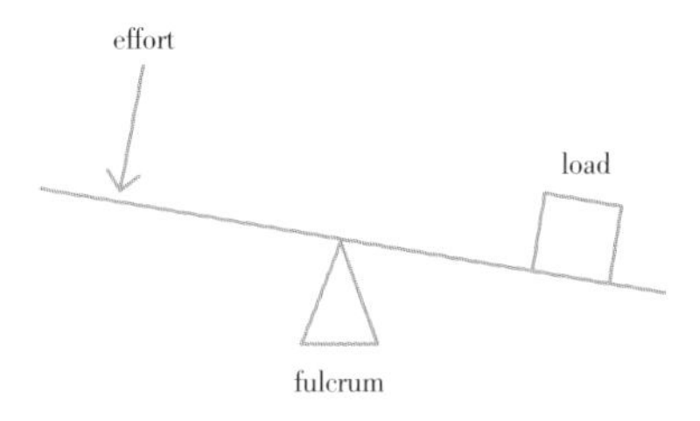


Each of these letters represents a value related to the lever:

- L1 is the value Archimedes is looking for—the length of the lever from the point where the force is applied to the fulcrum (the point at which the lever pivots).
- L2 refers to the length of the lever from the fulcrum to the world in inches. We know from the problem that this length is exactly 10,000 miles, which can be converted to 633,600,000 inches.
- m2 is the mass of the world, which has been estimated to be roughly $1.31695337 \times 10^{25}$ lbs.
- m1, finally, is the force which is being applied to the lever, which in this case is 150 lbs (assuming that Archimedes is merely standing on the lever and not applying any additional force).

## THE SOLUTION:

Now, all Archimedes must do is perform the calculations (which in this case can be a bit messy since there are all these extremely large numbers at work, so he must be very careful). He begins by plugging the known values into the equation and then solving for L1.

$$L1 = \frac{(1.31695337 \times 10^{25}) \times 633{,}600{,}000}{150}$$

Now, calculating all of these numbers (probably best to use a calculator for this one) provides a quick and easy answer to Archimedes' great problem:

$$L1 = 5.56 \times 10^{31} \text{ inches}$$

This is a huge number! Even after converting this number back into miles we find that the lever would have to be $8.78 \times 10^{26}$ miles long! In fact, it turns out that this distance is larger than the size of the known universe.

Although Archimedes may not be able to move the world with his lever after all, his point is pertinent—simple machines allow more work to be done using less energy.

# SCIENCE IN THE DARK AGES

By the second century BC, the hegemony enjoyed by ancient Greek civilization had waned. This was followed by more than 1,500 years of scientific stagnation in the western world. During this period, many of the great works of antiquity were neglected and eventually lost forever. Events such as the repeated destruction of the Library of Alexandria (which had to be rebuilt multiple times) during the first few centuries AD, compounded this. It is quite fortunate, then, that elsewhere in the world, there survive faint remnants of these ancient ideas.

**The Fall of the West**

When the Western Roman Empire fell, not only had Democritus's theory of tiny hypothetical particles (atoms) been long since discarded, but even the legacy of Aristotle, the greatest thinker of them all, found itself in danger of being forgotten entirely. Though the Roman world was not entirely without its scientists (Ptolemy, for example, in the second century AD was considered one of the finest astronomers of his day, offering up an incorrect, though still rather brilliant Earth-centered model of the solar system), the Western world (Europe) began its steady decline into that period known as the Middle Ages (a term generally assigned to the thousand or so years from the fall of the Western Roman Empire in the fifth century to the "Renaissance" period, which began somewhere around the sixteenth century). During this time questions of science fell off the lists of priorities.

**Alhazen and the Great Thinkers of the Middle East**

It is fortunate for the modern world that within the growing Arab communities of Persia and the Middle East a few notable figures were able to shed some scientific light on these "dark ages." The eleventh-century scholar Abu Ali al-Hasan ibn al-Hasan ibn al-Haytham (commonly referred to as Alhazen in the West), for example, possessed a particularly brilliant scientific mind. He performed some formidable studies on the subject of light—a phenomenon that had been even more elusive than atoms to the ancient thinkers.

Alhazen is often known today as "the father of optics." He was one of the first thinkers to utilize a truly inductive scientific method (wherein one begins with experiment and develops theories based on these observations) and to perform extensive experiments in this previously unknown field. Alhazen performed experiments using lenses and mirrors to produce some of the first calculations of light reflection and refraction. He also explored the separation of light into its constituent colors using a prism.

Although he is remembered in large part for his work in optics, during his career Alhazen also performed exciting work (which would escape the notice of the Western world for several more centuries) in mathematics, astronomy, medicine, and even psychology.

### Other Arab Scholars of the Dark Ages

While Alhazen remains arguably the most studied Middle-Eastern scientist of the Dark Ages, he was not alone. Other thinkers rose up, such as Ali ibn Isa, who, in the tenth century, began to explore the optical part of the brain and made a highly accurate measurement of the Earth's circumference, and Al-Battani, who in the ninth century made astronomical corrections to Ptolemy's solar and lunar tables and made numerous contributions to mathematics.

The Arab and Persian scholars of the ninth, tenth, and eleventh centuries performed excellent work in the realm of science, but perhaps the most important of all the Arabic contributions to the future of science was their devotion to saving and studying the extant works of Greek philosophy, especially Aristotle (see pp. 18–19), which they translated into Arabic, thus preserving them for the benefit of the entire world, allowing at least some of these remarkable volumes to be reintroduced into Europe centuries later, after the relative scientific stagnation of the Dark Ages.

### ALHAZEN'S *BOOK OF OPTICS*

Though Alhazen is credited as being the author of more than 200 scientific and mathematical works (about half of which have survived), one stands out as his most memorable: *Book of Optics*. In this work, Alhazen describes optical phenomena in greater detail than had ever been accomplished before. Over the course of seven books he looked at how light travels, how images are perceived by the eyes, how light can be altered and changed by way of reflection, refraction, or diffraction, and much more. It is truly one of the greatest scientific works of all time.

• An etching of Alhazen, as he was known in the West.

# A REVOLUTION IN OPTICS

It is remarkably easy to take pictures today. Since the advent of digital photography, there really is nothing to it. Just point, click, review, delete, and repeat the steps all over again *ad nauseum*. The entire process takes only seconds and no expensive film is wasted. Looking back just a few years, before the takeover of digital technology, taking pictures was a much more time-consuming and expensive process. Imagine, then, what it must have been like 900 years ago, when the first ever camera was invented.

**The First Camera**

The first device used to capture images probably wouldn't even be recognizable as a camera today (and in truth, calling it a "camera" in the modern sense may be a bit misleading), but in truth, it works on a very similar principle. The English word "camera" comes, in fact, from the Latin term used to describe this very first piece of photographic equipment: *camera obscura*, or "dark room."

While performing experiments in the field of optics and light (a field which he revolutionized), the Persian scientist Alhazen (see pp. 26–27) created a remarkably simple, yet fascinating invention in the eleventh century: a small room, closed off to all light except for a tiny hole in the wall that allowed a cone of light to filter through and project images of the outside world onto the inner wall of the room, as if a snapshot had been taken!

To Alhazen, the images projected onto the wall of the camera obscura were used as a basis for further scientific inquiry. It wasn't until the sixteenth century that artists finally began to use the camera obscura as a tool for capturing images of the world, though the only way to truly do so was to place a piece of paper or canvas on the wall and to trace the image that appeared inside the dark room, creating photo-realistic representations of the outside world.

It wasn't until 1826 that a French inventor named Joseph-Nicéphore Niépce took the first photograph, employing the principles of the camera obscura and adding to it a knowledge of chemistry. In Niépce's invention, the image that arrived in the camera became embedded onto a pewter plate covered with various chemicals, thus reproducing the image. Thirty-five years later, in 1861, physicist James

Clerk Maxwell (see pp. 76–77) became the first person to take a color photograph using filters of three different colors and then superimposing them onto one another.

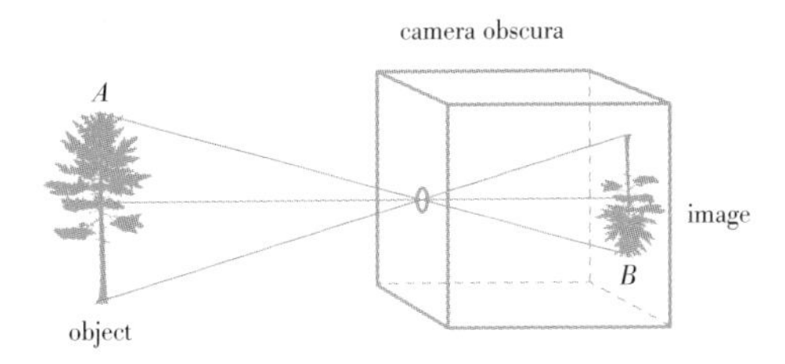


## Optical Principles in the Camera Obscura

Though it is hard to believe today, when Alhazen created his first camera obscura, very little was known about light. The use of this invention led to numerous discoveries regarding the nature of this rather elusive substance.

First, and perhaps most importantly, Alhazen discovered that light travels in a straight line (this is far more important than it sounds). Because of this, the images reproduced within a camera obscura appear upside down. As is demonstrated by the diagram on the right, any light coming from above continues to travel straight and arrives on a lower part of the wall, while light from below continues in a straight line and strikes higher up on the wall.

The camera obscura also became an important tool for observing the sun because it allows an observer to view an image of the sun, rather than looking directly at it. Through the camera obscura, solar eclipses could be observed without any danger to the eyes.

Though many of the principles of light remained a mystery long after Alhazen's time, his work, particularly through this remarkable device, represented a significant step forward. It allowed humankind to begin to peel away some of the layers of mystery from the phenomenon of light and to breathe new life into the world of science in the Middle Ages.

### BUILD YOUR OWN CAMERA OBSCURA

Many photographers and hobbyists today continue to experiment with photography by building their own "pinhole cameras," using the principles of the camera obscura as outlined by Alhazen more than 1,000 years ago. Consisting of just a box (a cereal box would be fine), a tiny pinhole, a homemade shutter, and a piece of photographic paper, there are numerous resources online and at local libraries which describe in simple terms how to construct these fun devices.

Exercise 3

# Optical Refraction

## THE PROBLEM:

Alhazen decides to take up a career as a spear fisherman in the Persian Gulf. Sailing out to sea he spots a shoal of fish several feet below the surface. Taking aim, he throws his spear but misses them. He attempts it numerous times but without success. Finally, he realizes that the problem is not his aim but light itself. How can Alhazen use his knowledge of optics to become a more successful fisherman?

## THE METHOD:

The problem being faced by Alhazen can be quickly and easily demonstrated with a simple experiment:

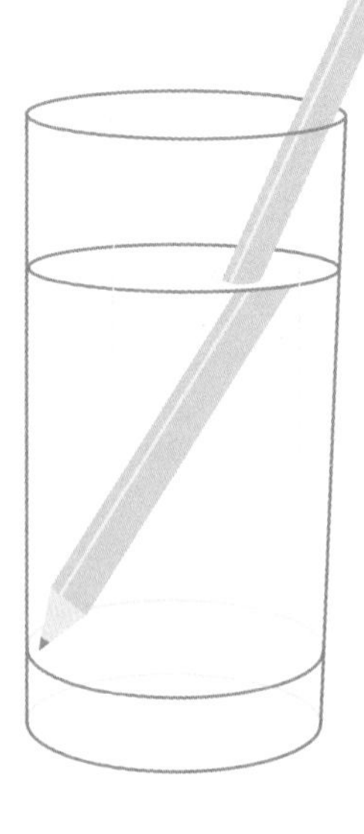

**1)** Take something long and narrow (a pencil will do quite nicely) and a clear glass filled with water.

**2)** Place the object into the water so that half is under water and half is sticking up out of the glass.

**3)** Take a good look at what is happening.

If the experiment is performed correctly, it becomes clear that the straight object is no longer straight! It appears to bend as soon as it enters the water! View the experiment from different directions and the angle will appear to change, as if the water itself was somehow causing the object to become physically deformed. But it's only water! What is going on here?

It wasn't until the first decade of the nineteenth century that the physicist Thomas Young finally gave the world a sufficient explanation of the phenomenon of light refraction. He demonstrated that light travels in the form of waves, much like the surface of water or the propagation of sound through the air. Finally, this phenomenon of refraction could be solved by recognizing that in refraction light appears to bend, though this is only an illusion brought about by its wave-like nature.

In modern optics, every translucent substance possesses what is called a refractive index, which is a measure of how much light slows down when traveling through it (which ultimately has to do with the chemical composition of the substance). When light waves pass through this substance (such as the water and glass in this experiment), the wave propagation speed is altered, while the wavelength remains the same. This means that, while the object may not change color or basic appearance, its relative position and shape will appear altered when viewed through a substance which has a refractive index different to that of air.

All of this is just a somewhat more complex way of saying that light bends as it passes through different substances.

## THE SOLUTION:

Though he does not understand that this phenomenon is due to the wave properties of light, Alhazen understands perfectly well what is going on here—for he was the first to calculate the refraction of light. He knows that the only way he will ever be able to spear any fish is by first taking into account the refraction of light and then adjusting his aim to compensate for it.

By noting the angle at which the light from the Sun is striking the water, calculating the angle of its refraction and the depth at which the fish are swimming, Alhazen adjusts his aim so that he is throwing his spear to the side of the fish, and just like that, he finds success! Suddenly, the laws of physics are not simply theoretical—they serve a fully practical purpose!

***Turn to pp. 84–85 for more information on light and optics.***

Chapter

# The Renaissance Begins

Although the Renaissance is more often associated with the arts, and artists such as da Vinci and Michelangelo, it was a period of significant progress in the sciences as well. This chapter introduces the groundbreaking theories of Copernicus, Galileo, and others—theories that helped instigate the study of science as we know it today. By asking new questions they began to offer real answers and in doing so developed a scientific method that remains influential to this day.

Profile

# Nicolaus Copernicus

**Nicolaus Copernicus laid the foundations for a revolution in scientific thought and ushered in a new era of scientific thought with his work, *De Revolutionibus Orbium Coelestium* (*On the Revolutions of the Heavenly Spheres*). By asserting that the planets revolve around the Sun, Copernicus brought about a true revolution in European science. It is for this reason that Copernicus is often credited as being one of the first to attempt to bring science out of the Dark Ages and into the Renaissance period.**

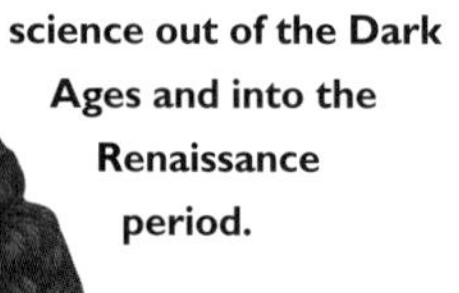

## The Monastic Astronomer

Copernicus is popularly held to be the founder of modern astronomy. He was born in Poland on February 19, 1473, and, after his father's death, was raised by his uncle, Lucas Watzenrode. Though he never married or had children, Copernicus lived a long life devoted to religion and his various and diverse scientific studies.

After studying mathematics and optics at Kraków University, Copernicus moved on to Bologna, Italy, where he studied canon law. His schooling complete, Copernicus moved again, this time to the Prince-Bishopric of Warmia, an ecclesiastical state in Germany, where he worked for his uncle in an ecclesiastic and administrative role.

Though he had many interests (including art and languages such as Greek and Latin), Copernicus soon became heavily interested in astronomy, making his observations (by way of

"At rest, however, in the middle of everything is the Sun."—***Copernicus***

**Key works**

- **On the Revolutions of the Heavenly Spheres**
  *Published in 1543, this work represents a defining moment in the history of science, for it is the first major work published which argued strongly for a heliocentric model of the Solar System. He not only explained, piece by piece over the course of six books, the revolutions of each of the planets and the movements of the Sun and Moon; he also provided detailed explanations of how the positions of astronomical objects can be calculated. This enabled his theory to be verified by way of prediction and experimentation—a radical departure from the past. It was the first truly "scientific" theory of astronomy.*

the naked eye, as the telescope had yet to be invented) from a turret on the wall surrounding his home.

Copernicus died in Frauenburg on May 24 1543. A frequently repeated legend has it that his most famous and revolutionary work, *On the Revolutions of the Heavenly Spheres*, was published on his final day, with the completed work presented to him only just before he died.

## The Revolutionary Idea about Revolutions

Though his final theory was not published until the time of his death, Copernicus had been pondering a heliocentric (Sun-centered) theory for some time. In fact, as early as 1514, almost 30 years before his death, Copernicus had written an unreleased work known as *Commentariolus* ("Little Commentary"), which he gave to friends and colleagues but never published during his lifetime.

In this work and others, Copernicus duly laid out the groundwork for the theory which would transform astronomy forever. Although it did not immediately supplant the reigning Ptolemaic System (that is, the Earth-centered system refined by Ptolemy in the second century), it provided a foundation upon which further thinkers such as Johannes Kepler (see pp. 36–37) and Galileo Galilei (see pp. 42–43) could further build on the theory.

Copernicus laid out his heliocentric theory based on a system of "spheres"—that is, orbits, born out of a tremendous number of precise calculations compiled by Copernicus. (These calculations were published after his death by astronomer Erasmus Reinhold in 1551, becoming the standard reference guide for astronomical tables and supplanting all previous such calculations.) To Copernicus, the known universe was comprised of eight spheres. Six of these represented the orbits of the planets known at the time—Mercury, Venus, Earth, Mars, Jupiter, and Saturn—and had the Sun at their center. The seventh was the outermost sphere (the "firmament") upon which the distant stars sat. The remaining sphere had the Earth as its central point, upon which the Moon was positioned.

The theory of the motion of the heavenly bodies and their revolution around the Sun marked a dramatic change of perspective that has had wide-ranging implications, scientific and otherwise. By basing his understanding of the universe solely on the principles of observation and mathematics, Copernicus established himself as a founding figure of "modern" scientific method.

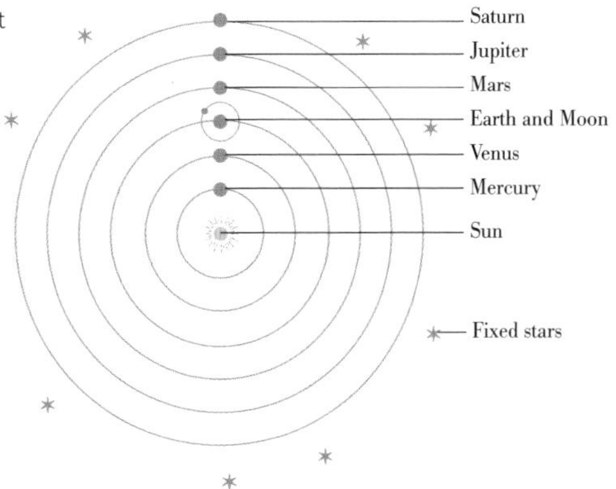


• The Copernican model of our Solar System consisted of eight spheres: six for the known planets orbiting the Sun, one for the Moon orbiting Earth, and one final sphere for the distant stars.

**Profile**

# Tycho Brahe and Johannes Kepler

**Tycho Brahe and Johannes Kepler were brilliant men in their own right. Brahe had an insatiable desire to better understand the motions of the objects in space, which led him to take some of the most important measurements of planetary motions, while Kepler had a unique mathematical mind which could interpret tables of data in order to find consistent patterns in the midst of seeming chaos. Together, they ushered in a new era of astronomy and physics.**

## Tycho Brahe: The Noseless Dane

The Danish nobleman Tycho Brahe (1546–1601) became enthralled with the idea of astronomy from a very young age, prompted by the fact that it had been possible for astronomers, even as far back as the sixth century BC, to not only observe but to predict events in the night sky, although even then he recognized that because of imperfect and incomplete astronomical data, these predictions could be off by a considerable margin.

Brahe set out to correct the errors in astronomy by compiling the greatest wealth of observational data ever seen, but only after a rather raucous young-adulthood, which saw him study law, travel extensively, lose his nose in a duel (he famously sported a false nose made of gold, silver, and wax), hire a clairvoyant dwarf (whom he kept under his dinner table), and domesticate a moose (that later died after drinking too much beer and falling down a set of stairs).

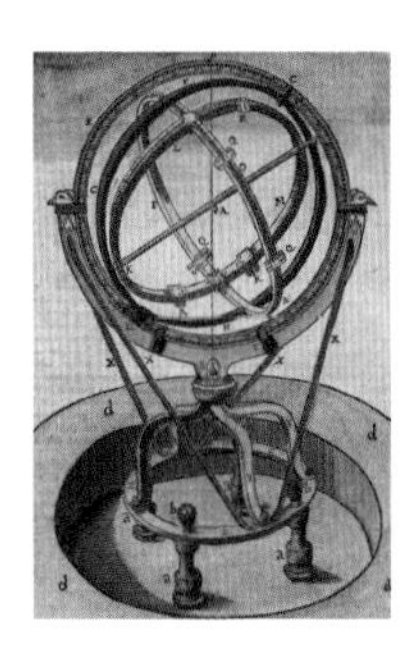

• Tycho Brahe's Great Equatorial Armillary was a large-scale model of the Solar System's "celestial spheres."

After discovering a new star and publishing a book on the discovery, Tycho was given the island of Hven (now Ven, Sweden) by the Danish King, Frederick II, to build the world's foremost astronomical observatory, called Uraniborg. He created large instruments—his "great equatorial armilla" (a model of the celestial spheres) was some nine feet (2.75 m) across; his "great mural quadrant" (an instrument used to measure the exact spatial location of heavenly bodies) was 13 feet (4 m) across. After a disagreement with Frederick II, Brahe moved to Prague, where he was appointed as imperial astronomer.

Tycho Brahe's legacy was not immediately apparent. For starters, he never came to accept the Copernican model of a heliocentric universe, instead choosing to restore Earth to the center and interpret his massive quantities of data using this model. It remained to be seen whether the data could support this view of the universe.

## Johannes Kepler: The Blind Astronomer

Johannes Kepler was born in 1571, the son of a mercenary soldier and an innkeeper's daughter (who, it is said, was tried for witchcraft). As a small, fragile man with poor vision, Kepler seemed unlikely to make much of himself, let alone to become a famous astronomer. Yet, thanks to his remarkable intelligence, Kepler won a scholarship, allowing him to attend the University of Tübingen, Germany.

Kepler's early interest in astronomy came as he pondered the Copernican astronomical tables and then sought to combine them with the geometry he was learning in school. Kepler followed mathematical clues to independently devise his own model where the orbits of the six known planets orbited along six spheres, each of which could be transcribed within one of Plato's Perfect Solids, from an octahedron within Mercury's orbit to a cube within Saturn's orbit (see p. 17).

Although Kepler's theory turned out to be wrong, the model impressed Tycho Brahe enough that he invited the young man to work with him. Eighteen months later, Brahe died of an acute urinary infection. His last words to his assistant, reportedly, were, "Let me not seem to have lived in vain."

Kepler was named Brahe's successor as the imperial mathematician and given the responsibility of completing the unfinished work of Tycho Brahe. His mathematical acumen allowed him to better understand and calculate the orbits of the various planets, leading to the discovery that the planetary orbits are not in fact perfect circles but ellipses.

Further research led Kepler to establish three laws of planetary orbits that would fundamentally change the science of astronomy, inspiring future generations of scientists, including Galileo Galilei (see pp. 42–43) and Isaac Newton (see pp. 54–55).

**Key works**

- **MYSTERIUM COSMOGRAPHICUM** *(1596): In his first major astronomical work,* The Cosmographic Mystery, *Kepler outlines his geometrical theory of planetary orbits.*
- **ASTRONOMIAE PARS OPTICA** *(1604):* The Optical Part of Astronomy *is one of the founding works in the field of optics. He also applied his understanding of optics to the human eye.*
- **ASTRONOMIA NOVA** *(1609): In* New Astronomy, *Kepler lays out the laws of planetary motion, including the conclusion that planets move along elliptical orbits.*

# KEPLER'S LAWS OF PLANETARY MOTION

Johannes Kepler's greatest achievement was to develop three succinct laws that defined the movement of every known astronomical body in the Solar System—laws still used by astronomers and physicists today, four centuries after their discovery. The three laws are remarkably simple.

### The First Law

*The orbits of the planets are elliptical, with the Sun as the focus of each one.*
What is the difference between a circle and an ellipse? Well, not much. Unlike a circle, which has only one point at its center, an ellipse has two (for more details on this, see the following page for an exercise on how to create and study ellipses). In essence, Kepler realized that the orbital paths of the planets were not perfect circles, with the Sun as a sole point of focus, but rather ellipses, with the Sun as one of two points of focus. What is at the other point? It is, in a sense, the cumulative gravitational attraction of everything else in the Solar System, which combines to turn circular orbits into ellipses.

Because of the elliptical nature of orbits, the distance between the Sun and the planets (or between Earth and the Moon) is always changing. There are points along every orbit where the object, or body, is closest to the Sun (this is called the "perihelion") and farthest from the Sun (this is called the "aphelion").

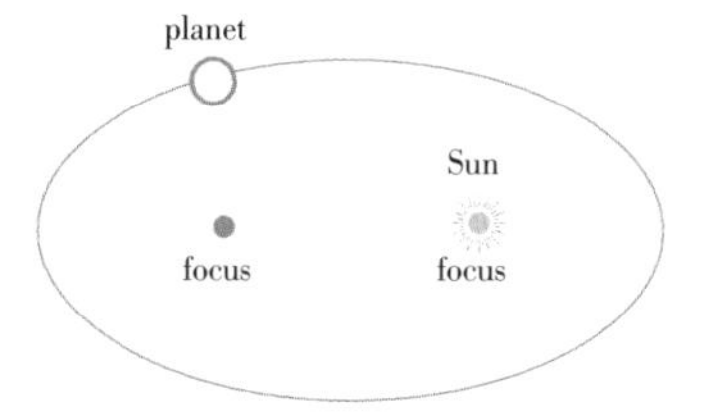


### The Second Law

*The line joining the planet to the Sun sweeps out equal areas in equal times as the planet moves along its orbit.*
This second law requires an element of decoding, although the principle it teaches is beautiful in its simplicity. Space agencies such as NASA use this law to predict the position of a spacecraft at any given moment.

Essentially, this law states that because the line joining the Sun and planet sweeps out equal areas in equal times, the planet moves faster when it is nearer the Sun. Thus, not only is the distance of a planet constantly fluctuating as it orbits the Sun, but also its speed as well. Therefore, the perihelion represents the fastest speed of an orbit, while the aphelion represents the slowest.

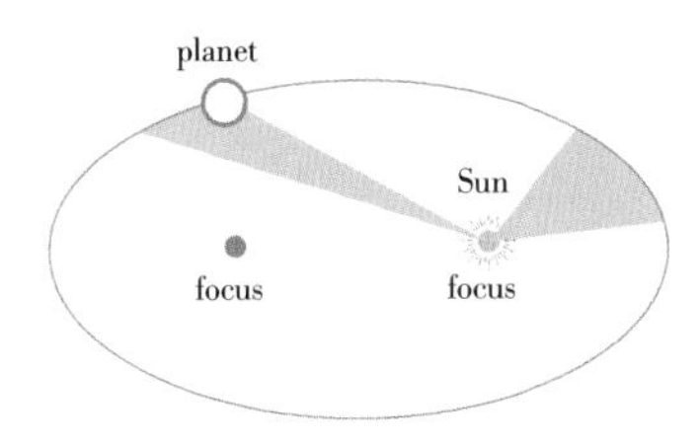


## The Third Law

*The square of the revolutionary period of a planet is equal to the cube of the semi-major axes of its orbit.* Probably the most complex of Kepler's laws, even this principle is not particularly difficult to digest. Perhaps it is even simpler when written in the mathematical form:

$$T^2 \propto O^3$$

In this equation (one of the simplest in all of physics) T represents the period of revolution for a planet and O represents the length of its semi-major axis.

Kepler's third law states simply that the orbital period for a planet around the Sun (which on Earth is 365 days) increases dramatically with the radius of its orbit.

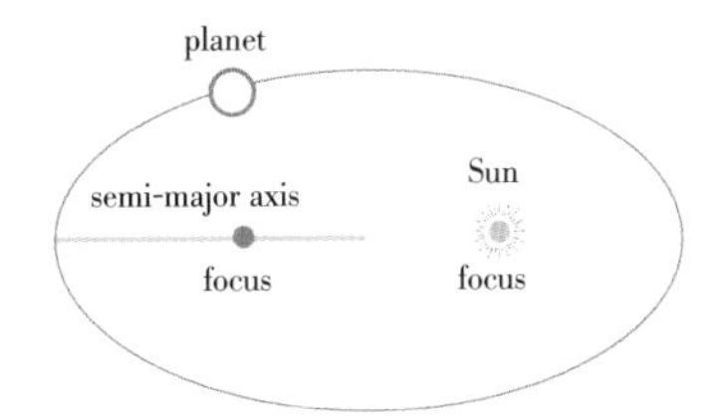


### PLANETARY ORBIT TABLE

| | Mean Distance from the Sun | Orbital Period (in Earth Days) |
|---|---|---|
| Mercury | 36,000,000 miles | 87.96 Days |
| Venus | 67,110,000 miles | 224.68 Days |
| Earth | 93,000,000 miles | 365.26 Days |
| Mars | 141,700,000 miles | 686.98 Days |
| Jupiter | 484,400,000 miles | 4,329 Days |
| Saturn | 887,900,000 miles | 10,751 Days |
| Uranus | 1,784,000,000 miles | 30,685 Days |
| Neptune | 2,801,000,000 miles | 60,155 Days |

# Demonstrating Elliptical Orbits

## THE PROBLEM:

As we've seen, Kepler surmised that the orbit of every planet is an ellipse, with the Sun as its central focus. He also knew that there existed a second focus within the ellipse. Next, Kepler worked to find a simple way of drawing an ellipse which demonstrated the effects of the two foci (plural of focus) on the orbits of the planets and moons. How did he manage this and what physical meaning could be drawn from it?

## THE METHOD:

Kepler found that the answer to his problem lay in just a few very simple tools: a couple of pushpins and a piece of string.

He experimented with these objects by pushing the two pins into a soft surface (such as a corkboard) a couple of inches apart. These represent the foci of the ellipse. Next, he cut the string to seven inches and tied it into a loop (the string must be long enough to make a loop which will fit easily around the two thumbtacks with plenty of slack).

Now Kepler could create his own ellipse. He took a pencil and used it to stretch the string tight, then he moved the pencil in a circular pattern around the two pins, using the string as his guide. Once he was finished, he had drawn a nearly perfect ellipse—a model of how planetary orbits behave!

But the experiment was not over just yet. Kepler, a true scientist, wanted to experiment further by adjusting the distances between the two pins, repeating the experiment and recording what happened to the shape of the ellipse as the distance between the two points changed. Kepler then asked the following questions:

- What factors are involved in the determination of the eccentricity (the degree to which it differs from a circle) of an ellipse? What must be changed in order to create a perfect circle?
- Why do planets seem to move in elliptical orbits rather than perfect circles? How do these pins relate to actual planetary orbits?

## THE SOLUTION:

The answer to Kepler's first question simply involves looking at how the shape of the ellipse changes with respect to the location of the pins and then drawing some general conclusions from several experiments.

Kepler noticed that the eccentricity of the ellipse increased as the pins were moved further away from each other and decreased as they were moved closer together. This means that as the two foci get closer and closer, the ellipse becomes more and more circular. In order to make a perfect circle then, he needs to place the two foci in precisely the same location (in other words, remove one pin from the board altogether).

This leads to the answer to the second question: Planetary orbits are elliptical because of the presence of two foci. One focus is, of course, the Sun. The other focus is a little more mysterious because, at first glance at least, it would appear to be nothing at all.

This is perhaps the most difficult thing to fully understand about basic planetary orbits. They are elliptical because they are gravitationally attracted to two points of focus, but one of these points does not even really exist! The reason, Kepler understood, is that while the Sun provides the vast majority of the gravitational influence upon the planets, they are also attracted to other things—in fact, every bit of matter in the universe is in some sense gravitationally attracted to every other! The other focus, therefore, can be seen as the cumulative attraction of all of these other objects (planets, stars, and galaxies, for example). Because this attraction is so weak compared to that of the Sun, however, most orbits in our Solar System resemble a model made with two thumbtacks placed very close together—very nearly circles, but not quite.

• A perfect ellipse can quickly and easily be drawn using just a pencil, a loop of string, and a couple tacks to act as the two focal points.

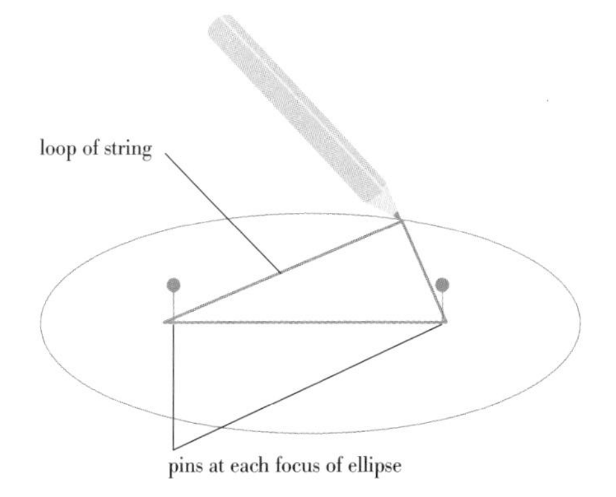

Profile

# Galileo Galilei

**Galileo perhaps best exemplifies the contrast between the science practiced by the ancient scholars and those who partook in the bountiful scientific achievements of the Renaissance period in Europe. Galileo is known for a great many scientific achievements, including the further popularization of the Copernican model of the cosmos, a scientific explanation of the pendulum, discovering the moons of Jupiter, dropping objects out of a tower, improving upon the design of the telescope, and helping develop one of the first real "scientific methods."**

## The Famous Martyrdom of Galileo

Most people surely know far more of the end of Galileo's life than the beginning. The story is oft-repeated, how he kindled the ire of the Roman Catholic Church with his insistence upon the theory that the Earth revolves around the Sun, rather than the other way around. Many have heard of the threats made against him, and how he was forced to recant those views he so eloquently laid out in his brilliant work, *Dialogue Concerning Two Chief World Systems,* and how, in a rousing demonstration of indignation, he muttered under his breath after recanting, "And yet, it moves."

Though these final words, like several other instances in Galileo's life, may be apocryphal, still they represent the crux of what is well-known of Galileo—that the final years of his life were lived out as a martyr to the cause of science.

This is all true enough, but to focus too intently upon this is to overlook a life absolutely filled with remarkable discoveries—so many that it is quite impossible to give a full account of them here.

## Notable Achievements

Born in 1564, in Pisa, Italy, Galileo is what today may be called a true Renaissance man, for he did not limit himself to one particular field of study but instead excelled at a great many. He was, like his father, a musician (he played the lute and made great strides in discovering the secrets of musical theory), and painter. He also studied medicine at a young age.

Galileo's accomplishments in science, of course, are many. He discovered the secrets of the pendulum's swing (see pp. 46–47) and, using primitive timing methods (including his pulse and the oscillation of a pendulum), he began performing experiments in motion, culminating in him proving once and for all that bodies of different weights fall at the same speed. (Though probably not by dropping objects out of the Leaning Tower of Pisa, as appealing as it may be—instead, he rolled balls down an inclined plane.)

In 1609 Galileo was first made aware of a new instrument invented by Hans Lippershey in the Netherlands called the telescope. By August of that year Galileo had not only built one of these instruments on his own, but had improved upon it tremendously. He went on to build some of the best telescopes in the world, which he used to view areas of space that no eye had seen. Soon enough, he had discovered moons orbiting Jupiter, the phases of Venus, the rings of Saturn and much more.

Galileo's infamous encounter with the Roman Catholic Church did not occur until he was nearly 70 years old and had already lived a full life of tremendous scientific achievement. It was with the publication of his *Dialogue Concerning Two Chief World Systems* in 1632 that Galileo courted trouble (even though Pope Urban VIII had personally invited Galileo's arguments for and against this system). By 1633 Galileo was ordered to stand trial for heresy, leading him to recant his views. For the last nine years of his life, Galileo lived under house arrest at his home near Florence, where he died in 1642.

• The first telescopes magnified distant objects only up to three or four times.

• Though Galileo did not invent the telescope, he certainly took advantage of the new technology by improving and popularizing it.

## Key works

• **Starry Messenger** *(1610): This work describes some of the first ever astronomical discoveries made with a telescope, describing mountains on the Moon, the Milky Way, and the moons of Jupiter. He also identified the Orion constellation.*

• **Dialogue Concerning Two Chief World Systems** *(1632): Formed as an argument between two men, Salviati (who believes in the heliocentric model of the Solar System) and Simplicio (who believed in a geocentric model), Galileo intended this revolutionary work to be an honest examination of the merits of both systems. The balance of the argument, which lay very much in favor of the heliocentric theory of the Sun as the center of the universe, is what landed Galileo in hot water with the Church.*

Exercise 5

# Falling Bodies Experiment

## THE PROBLEM:

For centuries before Galileo came along, the prevailing theory about motion and gravitation had been that of Aristotle. Perhaps most notably, the Aristotelian view was that the speed at which any two objects fall was directly related to their relative weights. Thus, a heavy object would fall faster than a lighter object. After all, pick up a feather and a rock and drop the two of them and Aristotle's view will prove sound: the rock will indeed fall faster than the feather.

The Aristotelian view on this matter appears to be self-evident, in a sense. As a result, no real attempts were made to test its validity for nearly two millennia. But of course, this view could not hold out forever.

Now, Galileo wanted to put Aristotle's idea to the test, but how does one even begin to perform an experiment on something that seems so intuitive? What methods did Galileo need to take in order to know once and for all whether or not objects truly do fall at different velocities depending on their masses?

## THE METHOD:

Galileo understood that these common sense notions of Aristotle might be nothing more than illusions. No, a feather does not fall as quickly as a rock, but this is because there is another force acting upon the feather that had not been accounted for: air resistance.

First, Galileo considered climbing up to the top of a building, such as the famous leaning tower in his home town of Pisa, and publicly dropping two objects to dramatically answer the question once and for all. But he knew that such an experiment is not exactly scientific, for there is still the chance that the objects could be affected by external forces, such as air resistance. The essence of a good scientific experiment is the element of total control.

So Galileo did not perform his dramatic public demonstration. Instead, he sought a method by which to measure force of gravity on objects while at the same time negating the effects of air resistance: he rolled balls of different weights down inclined planes.

## THE SOLUTION:

Inclined planes (ramps) serve to slow the acceleration of the balls just enough that the speed of descent can be accurately measured even with the crude time-keeping devices of the day (Galileo used either a measurement of his own pulse or the swinging of a pendulum to keep time).

But what did Galileo actually find by doing this? Simply put, he found that Aristotle was wrong. The balls rolled down the inclined plane at precisely the same speed, no matter how light or heavy they were.

Aristotle's theory paid no attention to air resistance. An object falling through the sky is resisted by the pressure of the air within Earth's atmosphere (such a thing would be no problem if the experiment were being performed on the Moon—a fact proven by American astronauts in the 1970s), and because the reaction of a body to force is proportional to its mass, lighter objects are affected more dramatically by air resistance than heavier objects.

There are many ways that we can demonstrate the truth of Galileo's "falling bodies" experiment today. For example, if we have access to a simple air pump, it is not terribly difficult to create a crude "vacuum tube" wherein the air within a chamber is sucked out, leaving us free to drop objects at will and observe that they all do fall at the same speed.

The work of Galileo shows that the force of gravity affects all objects on the Earth with equal force. It would be left to Isaac Newton a century later to calculate how this force actually works.

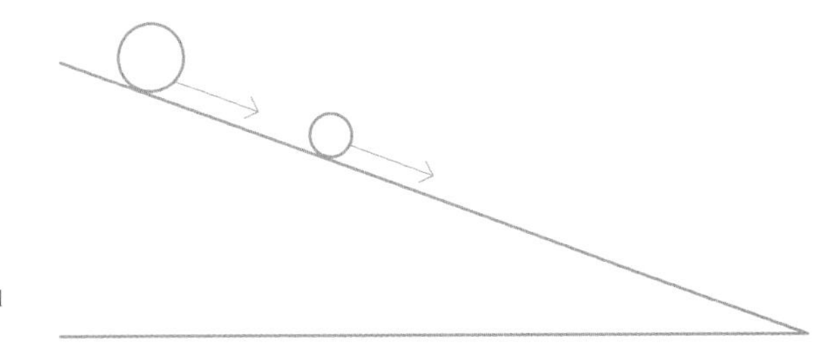

• Balls rolling down an inclined plane will move at the same speed regardless of weight.

# PENDULUM PHYSICS

Pendulums are remarkably simple devices, though they are not put into use as frequently today as they once were. The most common application for pendulums throughout history is in the mechanism of a clock, where one might see a narrow rod ending in a weight swinging back and forth, driving the complex gears found within. What is it that makes the pendulum such a useful item in keeping time? This was the question taken up by Galileo Galilei, who quickly found a rather simple formula to explain the behavior of the pendulum.

## Galileo's Remarkable Discovery

Pendulums are not limited to merely the metal pieces inside clocks. Any time a string with a weight at the end is swung back and forth, a pendulum is created. Whenever a hypnotist waves a watch in front of someone, he is utilizing the principle of a pendulum. A pendulum consists merely of a rod, wire, or string of fixed length from which a weight is suspended (*pendulus* in Latin means "hanging"). Its motion consists of the weight being allowed to swing freely back and forth, carried along a fixed path by gravity.

The time it takes for a pendulum to swing from the peak of one swing to the other is called the period of vibration.

What is it which determines the period of a pendulum's vibration? Is it its initial speed? Is it how hard it is initially pushed? Is it the mass of the weight? Galileo discovered that none of these answers are correct. In fact, the period of a pendulum is dependent on only one factor: its length.

As the story goes, Galileo was sitting in a cathedral, observing the swinging of a chandelier and, feeling inquisitive, timing its periods using his pulse. This simple experiment led Galileo to realize, to his great surprise, that the period remained ever constant—even as the chandelier's movement slowed its period remained the same!

Of course, a mind like Galileo's could not rest after the discovery of this peculiarity. He was soon experimenting with pendulums of various shapes, sizes, and lengths, and quickly discovered that the only aspect of a pendulum's construction which needed to be changed in order to determine its period was its length. Whether they are swinging in long arcs or have slowed down such that they are hardly swinging at all, the timing remains constant!

### Bringing the Pendulum from Theory into Practice

The consequences of this seemingly trivial discovery are actually quite remarkable. It meant that Galileo suddenly had access to a simple device which could allow him to keep time with much more accuracy than his own pulse! By building pendulums of specific lengths one could be sure that they would oscillate at a perfectly steady tempo. Within a few decades of Galileo's death, the Danish physicist Christiaan Huygens had developed an elegant mathematical equation which defined the period of the pendulum:

$$P = 2\pi\sqrt{(l/g)}$$

Here, the period (P) is defined in terms of the period's length ($l$) and the force of gravity ($g$). Using this simple formula, pendulums became the most accurate method of time keeping the world had ever seen, and would remain such until as late as the 1930s.

It was not long before pendulums had been partnered with systems of gears, wheels, and moving hands to produce the first truly accurate time-keeping devices.

• Foucault's invention uses the physical properties of pendulums to demonstrate the rotation of Earth around its axis.

## FOUCAULT'S PENDULUM AND GRAVITATION

There is another factor apart from length that affects the period of a pendulum's swing: the force of gravity (see pp. 46–47). The very thing which keeps a pendulum in motion also determines the speed and direction of its swing. In 1851, Leon Foucault created a massive pendulum which was free to swing in any direction, and as it did so traced lines in a bed of fine sand. After he had set it in motion, Foucault demonstrated that the pendulum had turned slowly in a clockwise direction. Since the vertical plane in which a pendulum swings does not change once the pendulum is set in motion, the tracings proved once and for all that Earth must be turning. This sort of device is now known as the Foucault Pendulum, and is readily on display at many science museums.

**Profile**

# René Descartes

**René Descartes is remembered today for his achievements in various fields, from pure sciences and mathematics to the most abstract philosophy. Though a contemporary of Galileo, Descartes was not so singularly focused on performing experiments and developing concrete scientific theories. His focus was on determining precisely how one should endeavor to think about things in the first place. He thought and wrote about logic, about method, and about how one might come to finally understand the universe in full.**

## Life as a Traveling Scholar

Descartes was born in 1596 and educated at the Jesuit college of La Flèche in Anjou. He entered the college at only eight years old, studying the basic (and mostly unquestioned) tenets of Aristotelian philosophy and essential mathematics. From a young age, Descartes suffered from poor health and from this he developed the life-long habit of remaining in bed until late morning each day.

After graduating from La Flèche in 1612, Descartes spent the next few years traveling restlessly, first to Paris and on to the University of Poitiers, where he received a law degree in 1616, then to Breda, Nether-

### Key works

• **Le Monde** *(1664):* The World *was written in 1628 but published posthumously in 1664. It is Descartes' first systematic presentation of his natural philosophy and includes an early description of atoms and a description of the Copernican model of the Solar System.*

• **Discours de la Méthode** *(1637):* Discourse on the Method *is the work in which Descartes sought to unify all scientific, mathematical, and philosophical topics by way of reason. The work set the benchmark for "modern" scientific practice.*

• **Meditationes de Prima Philosophia** *(1641):* Meditations on First Philosophy *describes Descartes dualistic philosophy, in which he argues that the human mind is separate from the body, and that our thoughts confirm our own reality.*

• **Principia Philosophiae** *(1644):* Principles of Philosophy *laid out Descartes' thoughts on natural philosophy, including his assertions that, free from external forces, the motion of an object will be constant and follow a straight line. Descartes intended this book to be a radical departure from the curriculum taught at the time in western European universities, particularly in Britain and France.*

lands, where he enlisted in military school. In the Netherlands Descartes studied mathematics extensively before setting off, from 1620 to 1628, on even more travels through Europe, including Bohemia, Hungary, Germany, Holland, and France.

By 1628 Descartes had grown weary of traveling, so he settled in Holland, where he wrote his first major work, a physics treatise called *Le Monde* (*The World*). In this book Descartes stated that he believed matter to be made up of tiny "corpuscles" (a precursor to modern atomic theory) and he described a universe in which Earth moved around the Sun. At that time, however, he heard of Galileo's troubles in Italy for espousing similar views, and consequently delayed publication of his work. It was not released until after Descartes' death.

## The Method of Science

Once started, Descartes' studies moved along rather quickly. His later work led him to tackle deep philosophical questions regarding truth and the nature of humanity from a purely logical perspective. Also a gifted mathematician, Descartes set about devising a comprehensive plan for combining science and mathematics to produce a more objective framework for the development of human knowledge. Though he did not deny the basic tenets of religion, Descartes believed that pure science and reason were all one needed in order to fully understand the workings of the universe.

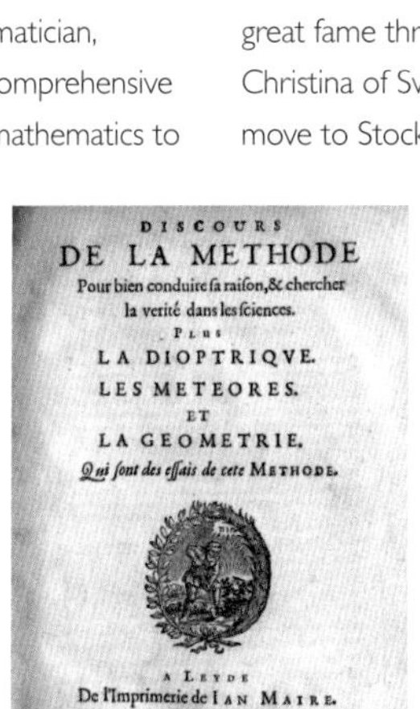
DISCOURS
DE LA METHODE
Pour bien conduire sa raison, & chercher
la verité dans les sciences.
PLUS
LA DIOPTRIQVE.
LES METEORES.
ET
LA GEOMETRIE.
*Qui sont des essais de cete* METHODE.

A LEYDE
De l'Imprimerie de IAN MAIRE.
cIↄ Iↄ c XXXVII.
*Avec Privilege.*

• Descartes' *Discourse on the Method* is the author's grand tribute to the power of human reason.

Perhaps the most important of Descartes' works is his *Discours de la méthode pour bien conduire sa Raison et chercher la Vérité dans les Sciences* (commonly referred to as *Discourse on the Method*), which he finally published in 1637. In this monumental work, Descartes began with simple principles and taught his reader how best to think about science, first by questioning everything (that is, removing all bias) and then building up a theory based on sound observational evidence. Descartes also used this work to invent one of the most common tools in all of mathematics: the Cartesian coordinate system.

A few years after publishing the *Discourse on the Method*, Descartes moved into more abstractly philosophical territory in *Meditationes de Prima Philosophia* (*Meditations on First Philosophy*), which was published in 1641, wherein Descartes offered his famous explanation as to how one might know that they exist: "I think, therefore I am."

Descartes' methods, whether philosophical, mathematical, or purely scientific, and his quest to unify science and mathematics, led him to great fame throughout Europe. In 1649 Queen Christina of Sweden persuaded Descartes to move to Stockholm in order to become her personal tutor. Unfortunately, this involved a dramatic change in the scientist's routine and saw him forced to wake up as early as 5am. After only a few months of the cold climate and early mornings, René Descartes died of pneumonia in 1650.

# THE SCIENTIFIC METHOD

During the Renaissance, European scholars were responsible for advancing human understanding of theories such as motion, gravitation, and optics, as well as forging new ground in areas such as mathematical physics. Perhaps the single most important of these, however, was the development of a clear, concise, and consistent scientific method—a system by which both observation and reason were employed in order to test more rigorously the theories of the universe.

**The Steps of the Scientific Method**
Although there is no singularly defined series of steps by which every scientist conducts their research, there is at least a generally agreed method which constitutes good practice. It tends to look something like this:

**1.** Make an observation about the universe: for example, a hiker walking through a forest spots two black bears.

**2.** Describe this observation and use it to develop a general rule, called a hypothesis, that is consistent with the observation: the hiker thinks to herself, "those bears are black, therefore all bears must be black."

**3.** Use the hypothesis to make predictions about future experiments: the hiker predicts that when she sees another bear, it will also be black, as will the bear following that one.

**4.** Test those predictions by performing experiments and making further observations. Be prepared to modify the hypothesis in the light of these latest results: the hiker sees that the next bear is brown—she amends her hypothesis to include brown bears as well as black bears.

"Science is organized common sense where many a beautiful theory was killed by an ugly fact."—***Thomas Henry Huxley***

**5.** Repeat the pattern of observation and refinement of hypothesis until there are no longer any discrepancies (the hiker travels all over the world, and has soon enough found black bears, brown bears, white bears, black-and-white bears, gray bears, and bears of a few other colors; eventually she has expanded her hypothesis to include every bear she could find on Earth). The greater the number of accurate observations, the more accurate the hypothesis. This is the foundation of the scientific method.

### Finding Truth in Science

There is no real last step to this basic scientific method, unfortunately. The nature of science is such that these steps must repeat endlessly, and the truth of the matter is that in science one really never can take the final step and declare something a "fact."

A theory may only be as complete as the experimentation that is used to develop it—thus, when we say something like "The force of gravity is felt between any two objects in the universe," what we have is only a theory—it is a generalized prediction based on what we have so far observed about our universe, but of course we have not actually gone out and tested this theory on every single object in the Solar System. If we did, perhaps we would find that somewhere in a distant galaxy there are objects which do not obey the law of gravity. If we did, we would immediately recognize that our theory was incomplete and we would have our work cut out trying to fix it.

On the surface, this may make the pursuit of science seem frustrating, but it is also one of the greatest things about it—the idea that at any time, any one of our most cherished theories of the universe might potentially be upset is extremely exciting. In fact, as we will learn in chapters five to seven of this book, this is precisely what happened in the beginning of the twentieth century.

### OCCAM'S RAZOR

Occam's razor is a logical principle attributed to the fourteenth-century friar William of Ockham that has become a hallmark of scientific theories ever since.

Occam stated that "Entities should not be multiplied unnecessarily." This has been taken to mean that wherever two or more explanations exist for a given phenomenon, the simpler one is generally preferable.

In other words, scientists use Occam's "razor" to cut away any superfluous elements of a theory, preferring to leave only the simplest explanation necessary, for logic dictates that the simpler answer tends to be preferable than an overly complicated one. This little rule is useful in any logical situation, even beyond the realm of science.

Chapter

# The Birth of Modern Physics

Beginning with the inestimable work of Isaac Newton toward the end of the seventeenth century, this chapter focuses on many of the essential elements of "Newtonian" physics. Newton's work laid the foundations for the physical sciences for well over 200 years and it continues to guide our understanding of the universe to this day. This chapter offers explanations on the theories of motion, gravitation, thermodynamics, and many of the other essential building blocks of physics.

Profile

# Isaac Newton

**Isaac Newton surely possessed one of the greatest minds in history. His achievements can hardly be overstated, whether they be in the pure sciences (the laws of force and gravity bear his name), mathematics (he invented calculus), theology (a lifelong passion), alchemy (perhaps his favorite subject) or any number of other interests. Newton's work represents a transition from the science of the Renaissance to a truly "modern" scientific approach.**

## Unusual Life and Personality

Isaac Newton was born prematurely on Christmas day 1642 in Woolsthorpe, England. His father had been a successful farmer, though he died before his son was born. After moving several times between the care of his mother and grandmother, Newton lived a relatively unhappy and tumultuous childhood which left him with a great many physical and emotional issues that stayed with him through life. He was famously difficult to get along with and emotionally unstable (suffering an emotional breakdown in 1678, the year of his mother's death), but his brilliance overshadowed all of this.

Newton finally left Woolsthorpe in June 1661 and entered Cambridge University, which would be his home and place of work for more than four decades.

In 1665, after Newton's graduation, Cambridge was closed due to the onset of the plague in England. Newton stayed at the family home during the two plague years of 1665 and 1666, and it was during this time that he was able to make some of his most memorable contributions, including the development of calculus, a revolutionary theory of light and

> "No great discovery was ever made without a bold guess."
>
> **—*Isaac Newton***

**Key works**

- **PHILOSOPHIAE NATURALIS PRINCIPIA MATHEMATICA** *(1687):* Mathematical Principles of Natural Philosophy *is Newton's quintessential work. It lays out Newton's laws of motion and his theory of universal gravitation. This is the work that defined physics for more than two centuries (and still has a great deal to teach us today).*

- **OPTICKS** *(1704): A remarkable work in which Newton described his revolutionary experiments in light, including reflection, refraction, his complex and visionary analysis of color, and his thoughts on prismatic dispersion. The end of this volume also deals briefly with Newton's theory of atoms, which was in itself quite revolutionary at the time.*

color and an attempt to explain planetary motion (which included the apocryphal "falling apple" incident which led to his insights on gravitation). The work done during these years eventually led to Newton's single greatest work, *Philosophiæ Naturalis Principia Mathematica* (*Mathematical Principles of Natural Philosophy*), which he published 20 years later in 1687.

Newton returned to the faculty of Cambridge in April 1667 and was soon elected a fellow at Trinity College, eventually becoming the Lucasian Professor of Mathematics (a post later held by such prominent scientists as Paul Dirac and Stephen Hawking).

Newton later moved to London in 1696 to take up the post of Warden of the Mint, where he achieved further successes. In 1703, Newton was elected president of the Royal Society (the premier society for scientists and intellectuals in England) and served in this position until his death.

In 1704 Newton published *Opticks* (which he had refused to release until after Robert Hooke had died, for Hooke had previously accused Newton of plagiarizing his gravitational theories). In 1705 he received a knighthood from Queen Anne. Newton died in 1727 and was buried in Westminster Abbey.

## Scientific Achievements

Isaac Newton is best known for his scientific achievements, which continue to impact almost all areas of science to this day. These include:

Mathematics—Newton formulated the binomial theorem of mathematics, as well as new methods for the expansion of infinite series. All of this was contained within the vast and complex branch of mathematics known as calculus. Calculus is at the heart of modern science and engineering—it has provided a platform for further discovery of how our universe works.

Optics—Newton took the study of optics to new heights, performing experiments using prisms which demonstrated that white light is composed of primary colors which can be separated from each other and recombined. He also reasoned (brilliantly but incorrectly) that light consisted of tiny "corpuscles" rather than as waves.

Mechanics—Newton's three laws of motion (see pp. 56–57) and his law of gravitational attraction (see pp. 60–61) became the cornerstone of all physics for the next 200 years. This is the foundation of what is even today known as Newtonian mechanics, which served as the preeminent theory of all physical interactions until the development of relativity and quantum mechanics in the twentieth century.

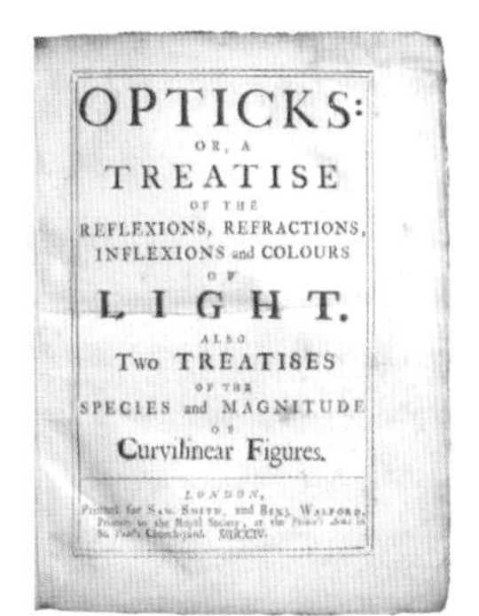
OPTICKS:
OR, A
TREATISE
OF THE
REFLEXIONS, REFRACTIONS,
INFLEXIONS and COLOURS
OF
LIGHT.
ALSO
Two TREATISES
OF THE
SPECIES and MAGNITUDE
OF
Curvilinear Figures.
LONDON,

• Newton's second great work, *Opticks*, gave the world the most thorough description of light yet developed.

# NEWTON'S LAWS OF MOTION

The formulation of the three laws of motion was perhaps Isaac Newton's greatest achievement. Even now, more than 300 years later, we are faced with their consequences every day. With every step we take, every action we perform, we see a demonstration of them.

### Newton's First Law

*A body at rest remains at rest and a body in motion remains in motion until it is acted upon by an outside force.*

Newton's first law is also called the law of inertia. Inertia is any body's tendency to resist a change in motion (it tells us how hard it is to get an object moving as well as how hard it is to stop an object once it starts moving).

The first half of this first principle seems obvious. No one would doubt that an object will not simply start moving on its own—it has to be acted upon by a force. A car will not simply roll down the street unless it is on an incline (in which case the force of gravity is acting on it) or it is being driven forward by its engine.

It is the second half of this law which is counterintuitive—an object in motion will continue in motion? Doesn't that seem to go against everything we observe? If I put an object in motion, eventually it always stops. Always. That is precisely what Aristotle and the ancient Greek philosophers observed when they declared the opposite law—that every object's most natural state is that of rest, and therefore every object will eventually come to rest. That, at least, fits quite well with experience. What Newton realized, however, was that an object's tendency to stop moving is not because nature wants it to stop, but because there are always forces acting against it, causing it to slow down. These forces come in many varieties; for example, friction, air resistance, and physical barriers, and they are all acting against every moving object on Earth. Only in the vacuum of space may we begin to see the truth of this law.

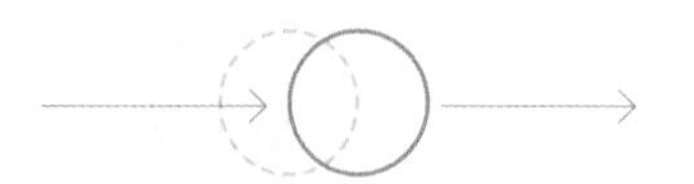

A rolling ball will continue at a constant speed and direction unless acted upon by another force.

### Newton's Second Law

*The force applied to a body produces a proportional acceleration; the relationship between the two is* $F = ma$.

In this simple equation F represents an applied force, *m* represents the mass of a body, and *a* is the body's acceleration. This law implies that there is a relationship between the force being applied to an object and the degree to which the object accelerates, but that it is directly dependent upon the mass of the object. Therefore, it will take more force to accelerate a heavier object than a lighter object (it is much more difficult, in other words, to push a truck down the road than it is to push a small Volkswagen Beetle).

The acceleration of an object is proportional to the force acting on it and inversely proportional to its mass.

Newton's achievement in stating this law was to offer the world a simple and precise equation explaining the relationship between force, mass, and acceleration. He showed that a double amount of force will double the acceleration; a triple force triples the acceleration. Furthermore, to calculate the motion of a body one needs to calculate all of the forces acting upon it, in every form and from every direction.

## Newton's Third Law

*To every action there is always an equal and opposite reaction.*

This law is often called the "law of reciprocal action." Imagine two people on roller skates pushing against each other; according to Newton's third law, the two skaters will each undergo equal changes of motion, but in opposite directions. What must be noted here, however, is that these two skaters will each be driven back by an equivalent force. This does not mean that they will both be driven backward at the same speed, for as we learned via the second law of motion, the acceleration of a body is directly dependent on its mass, so if the two skaters differ in mass even slightly, they will find themselves moving away from each other at different speeds.

The third law means that all forces are interactions, and thus that there is no such thing as a unidirectional force. You cannot push something away from you without also pushing yourself backward. A rocket cannot lift off Earth without also pushing Earth away from it. If a car starts moving forward, the tires are applying force to the road which is equal to the car's forward motion. Force is an interaction.

• Newton's cradle demonstrates the principles of the third law as well as the idea of the conservation of energy (see pp. 68–69).

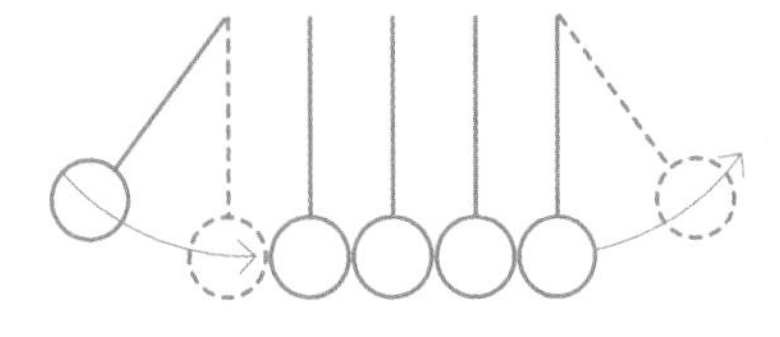

The left-hand ball decelerates from its initial speed to zero.

The right-hand ball accelerates from zero to the speed at which the left-hand ball was moving.

Exercise 6

# Calculating Velocity

## THE PROBLEM:

Two men are standing on a sheet of ice. The larger of the two men is a giant weighing 300 kg while the smaller man weighs only 100 kg. For reasons unknown, the smaller man reaches out and pushes the larger man with all his might—a total force of 10 Newtons. What will be the result of this action? Which of the two will slide across the ice, and with what velocity?

## THE METHOD:

Look close enough and you will find that all three of Newton's laws are required in order to fully understand the physics of this exercise.

**Law 1:** *A body at rest remains at rest and a body in motion remains in motion until it is acted upon by an outside force.* As soon as the small man pushes the larger man he is applying a force. Newton's law states that this will cause motion—clearly there will be some sliding involved.

**Law 2:** *The force applied to a body produces a proportional acceleration; the relationship between the two is* $F = ma$. From this law we know that we can calculate the exact acceleration produced by the application of force in this example. We already know the force (10 Newtons) and the mass of the two men (300 kg and 100 kg). This is enough to calculate the acceleration using the equation $F = ma$.

**Law 3:** *To every action there is always an equal and opposite reaction.*
The third law tells us that when the smaller man pushes the larger man they will both move, for the force acts upon them both equally.

## THE SOLUTION:

As we have seen, the three laws of motion dictate that the application of force in this scenario will result in the movement of both men. We can now apply the formula for force ($F = ma$) to calculate their respective accelerations.

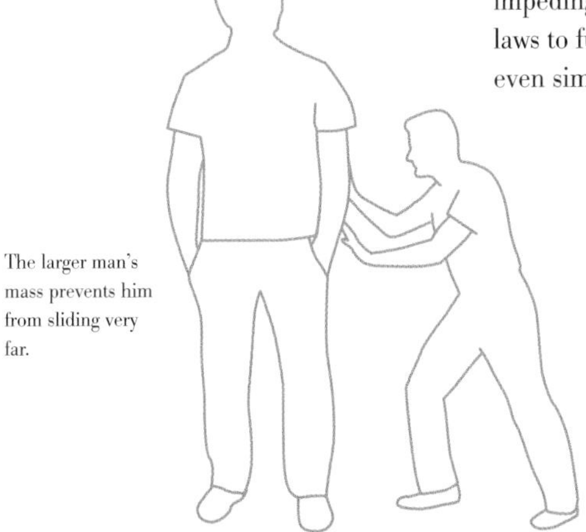

The same amount of force is applied to both individuals.

The larger man's mass prevents him from sliding very far.

The smaller man's inferior mass means that he will slide further.

Acceleration for the larger man:

$$10 \text{ N} = 300 \text{ kg} \times a$$
$$10/300 = a$$
$$a = 0.03 \text{ m/s}^2$$

Small man:

$$10 \text{ N} = 100 \text{ kg} \times a$$
$$10/100 = a$$
$$\text{a} = 0.10 \text{ m/s}^2$$

Though the force exerted on these two individuals is identical as per Newton's third law, their resulting acceleration is dependent on their respective masses, as per the second law. The large man slides backwards with an acceleration of just 0.03 $\text{m/s}^2$, while the smaller man will slide at an acceleration of 0.10 $\text{m/s}^2$. The force is identical, but the velocity is dependent on mass.

What's more, as per the first law, they will not continue to slide backward indefinitely, but will soon enough come to a stop as a result of the forces impeding their motion. It takes all three laws to fully understand the physics of even simple motion.

# FALLING APPLES AND CELESTIAL SPHERES: GRAVITATION

Isaac Newton was almost certainly not hit on the head by a falling apple. That said, there does seem to be a point when he observed the falling of an apple and this led him to begin pondering: if there is a force which draws an apple to the ground, how far does that force extend? Does it extend further into the sky? Into the atmosphere? Into space? Could it be that this very same force is capable of holding the moon in place around Earth? Such reasoning eventually led Newton to prove that this is indeed the case. Everything obeys the force of gravity.

**The Law of Gravitation**

Here's the mathematical formulation Newton came up with:

$$F = G\frac{(m1 \times m2)}{r^2}$$

This equation is used to measure the force of gravitational attraction between two bodies: F refers to the force of gravity; G refers to a universally constant measure of gravitational attraction (which hadn't been precisely measured at Newton's time, though he could offer a rough approximation); *m1* is the mass of the first object; *m2* is the mass of the second object; and *r* is the distance between them.

Today this is known as an inverse square law—an equation that defines a force which is inversely proportional to the square of the distance between two objects. As can be seen in the equation above, the factor most crucial in determining gravitational attraction is the distance between the two objects—as the distance increases,
so the gravitational attraction will decrease exponentially.

**The Accumulation of Gravity**

One thing that we rarely consider when looking at gravity is just how complex this force is. When we consider two bodies being gravitationally attracted to each other, we are really imagining two objects as if they were completely solid bodies—in truth, Newton's law is only simple when one thinks of the most basic objects: particles. In reality, the gravitational attraction between two large objects, such as Earth and the Moon, is a massive conglomeration of gravitational attractions between the

## MASS AND WEIGHT: WHAT'S THE DIFFERENCE?

An object's weight is dependent on gravity, so the weight of an object on the Earth will be different from the weight of an object on the Moon. Mass, on the other hand, is entirely independent of gravity—attempting to move an object on Earth should require precisely the same amount of energy as moving an object in space, even if it seems more difficult on Earth (because of other, opposing forces, such as gravity, friction, and air resistance).

many particles that make up Earth (including you and I) and the many particles which make up the Moon. When calculating Earth's gravity in a given location, it is easy to forget that objects are not simply attracted to one particular point at the center of Earth's core. Everything is attracted to every particle that makes up Earth.

Newton reasoned that if Earth is spherical (which it nearly is, though not perfectly so) then his equation works because all of these particles can be added together so that the planet acts as if there was but one "center of gravity."

### Gravity and Planetary Orbits

Imagine a cannon sitting on top of a mountain, firing horizontally. Eventually, a cannonball fired from this position will fall back to Earth. By adding more gunpowder, it is possible to extend the distance it travels before being dragged down by Earth's gravity. As still more powder is added, the ball will go even further, and its trajectory (the degree of its acceleration toward Earth) will decrease in steepness. Now, imagine that we put so much powder into the cannon that the ball's trajectory matched the curvature of Earth–the cannonball would not fall to the ground, but would circle all the way around Earth. It would be in orbit! Of course, in this example we are paying no consideration to air resistance or any other potential factors that might prohibit the possibility of this, and neither was Newton when he imagined that this was precisely how planetary orbits worked.

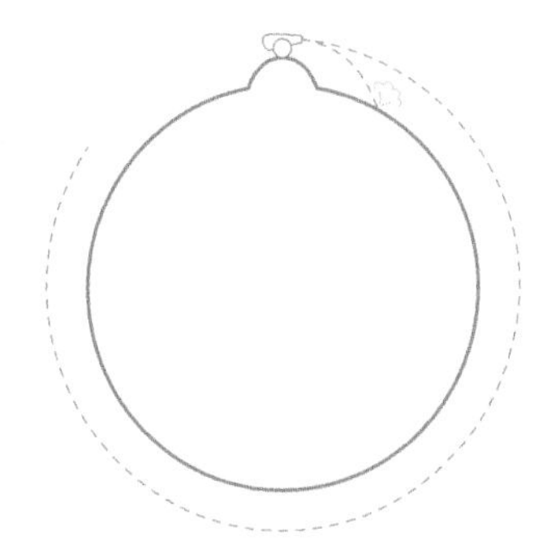

• Given enough gunpowder, Newton theorized that the trajectory of the cannonball would be such that it could orbit Earth.

# Measuring Acceleration

**THE PROBLEM:**

Isaac Newton spent the last years of his life attempting to understand the force of gravity and develop a method of calculating the gravitational attraction between two objects. Once he had achieved this, he attempted to put his famous equation to practical use by calculating the effects of gravity upon himself as he stood upon Earth's surface. How did he calculate this, and how would the effect of gravity change if he were to travel 1,000 kilometers into space?

**THE METHOD:**

Fortunately, Newton himself provided the essential mathematical tools necessary to answer these questions very simply. He begins by using the law of gravitation, a detailed explanation of which can be seen on the previous two pages.

$$F = G\frac{(m1 \times m2)}{r^2}$$

In order to solve this particular problem, Newton had to change the equation very slightly because he was no longer attempting to calculate the *force* of gravitational attraction between two distant objects in space (such as between Earth and the Moon). Rather, he wished to calculate the gravitational attraction of Earth itself upon any object on its surface. To do this he needed only to remove *m2* (mass of the lightest object) from the equation, leaving him with an even simpler equation:

$$F = \frac{GM}{r^2}$$

## THE SOLUTION:

Now all that is needed is to plug into this equation the correct values for each of these terms, as measured by other scientists:

*Mass of the Earth*: $5.97 \times 10^{24}$ kg
*Distance to the center of the Earth*: 6,378,100 meters (although this number changes depending on the exact location on Earth, as the planet is not a perfect sphere).
*Gravitational constant*: $6.67 \times 10^{-11}$ $Nm^2/kg^2$

These numbers can now be added and the equation solved to find the value of F:

$$F = (6.67 \times 10^{-11}) \times \frac{(5.97 \times 10^{24})}{6{,}378{,}100^2}$$

$$F = \frac{3.98199 \times 10^{14}}{6{,}378{,}100^2}$$

$$F = 9.79 \text{ N}$$

This may strike many readers as familiar. It is a commonly known figure that most students learn early on in their education: Earth's gravity pulls us towards it with a force of approximately 9.8 N. This force determines the maximum speed an object will reach if falling towards Earth.

Now, to quickly solve the second part of Newton's problem: how does this number change if you were to travel 1,000 kilometers into space?

$$F = (6.67 \times 10^{-11}) \times \frac{(5.97 \times 10^{24})}{7{,}378{,}100^2}$$

$$F = \frac{3.98199 \times 10^{14}}{7{,}378{,}100^2}$$

$$F = 7.32 \text{ N}$$

So, by traveling 1,000 kilometers into space, you would still very much be affected by Earth's gravity. As the distance from the surface of Earth increases, this number decreases. In fact, by the time we reach the surface of the Moon, the Earth has a relatively meagre gravitational pull on the Moon of 0.0073 N. These numbers decrease rapidly because gravity operates according to the "inverse square" law, which means that the strength of gravity is inversely proportional to the square of the distance between two objects. It never disappears entirely, however—no matter how far into space one travels.

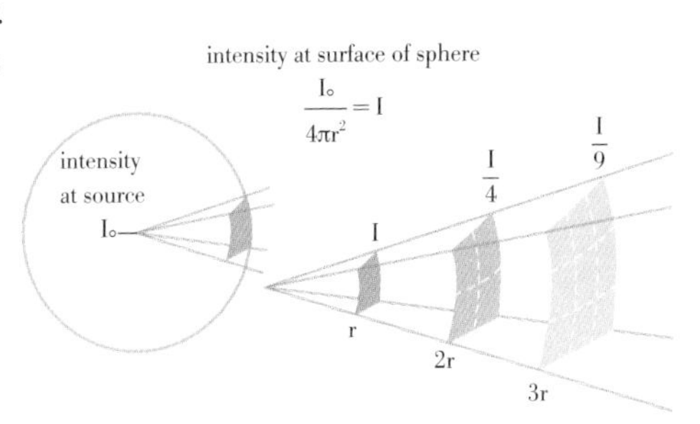


• Newton's theory of gravity suggested an "inverse square law," where the strength of the force diminishes rapidly as distance from the source increases.

# BOYLE'S LAW

Even before Isaac Newton, another branch of scientific research had already begun which would continue to be developed well into the nineteenth century—research into the nature and behavior of gases and other elements. Much of this work was performed by some of the first chemists and physicists, such as Robert Boyle and Daniel Bernoulli, whose exploration of the gaseous elements would lead to far greater understanding of chemistry, atomic theory, and even everyday mechanics.

### Discovery

Before his most important work was performed in the 1660s, Robert Boyle, an aspiring alchemist (a field which combines supernaturalism and science to seek the true nature of the elements) read of a new air pump invented by Otto von Guericke. This device allowed scientists to create a vacuum within an enclosed container. Boyle set about building one of these pumps for himself, which he then used to investigate the properties of various forms of gases. From this research came Boyle's law, an essential formula for understanding the nature of gases. This law states that for any given amount of gas at a constant temperature, its pressure and volume are inversely proportional to one another. This means that, as the volume increases, the pressure decreases by a proportional amount, and vice versa. This theory could be rather easily verified experimentally, but its deeper significance would not be understood for some time to come.

### The Mathematics

The mathematical equation for Boyle's law is:

$$pV = k$$

In this formula, $p$ represents the pressure of a system (usually represented in units called pascals, or atmospheres); V represents the volume of the gas (measured in cubic meters); and $k$ represents a constant value representative of the pressure and volume of the system.

Boyle's law, it should be noted, relates only to a situation in which the temperature is constant (for as Boyle also noted, the pressure of a gas can be altered by a change in the temperature). Under these conditions, Boyle's law states that the volume of a gas is inversely proportional to the pressure applied to it.

This means that in an inflated balloon, the space that the air occupies will vary in size depending on the pressure surrounding it. On the surface of Earth, the pressure is roughly 1 atmosphere (which

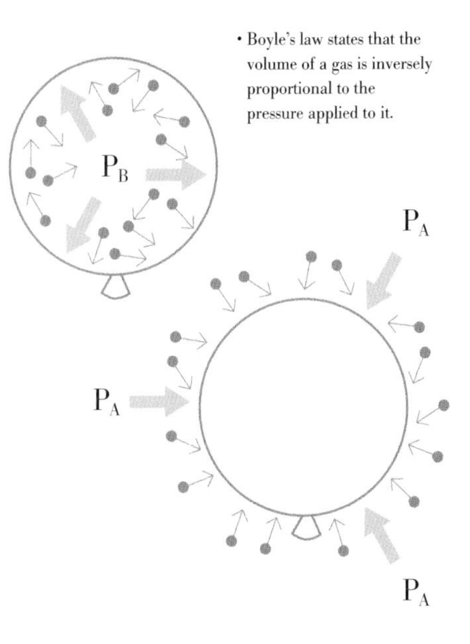


• Boyle's law states that the volume of a gas is inversely proportional to the pressure applied to it.

is the unit of atmospheric pressure, or atm for short). As the pressure reaches 2 atm the volume of the air within the balloon will shrink by half. Four atmospheres will reduce the balloon to one quarter of its original volume.

## Boyle's Law in Action

Boyle's law has many everyday applications, which you may have overlooked up until now. When the plunger is pulled out on a syringe, for example, the volume within the cylinder is increased, thus decreasing the pressure. This pressure then draws in liquid (such as blood) or gas to fill vacuum. In addition, the human respiratory system would not operate if Boyle's law was not true, and anyone who has ever successfully built a submarine has made note of the effects of Boyle's law. As soon as the pressure outside the body of the craft becomes great enough and the gases within become sufficiently pressurized, there lies the possibility of collapse.

Boyle's law is arguably more important today than it was when Boyle formulated it in 1662. It lies behind air, space, and undersea travel, as well as behind many other applications.

### THE BENDS

Though its effects were first recognized in bridge-builders working in pressurized chambers underwater, the devastating illness known as "The Bends" is perhaps most well-known among scuba divers, and is a perfect (though tragic) example of Boyle's Law in action. Imagine a diver descending into the depths of the ocean, breathing pressurized air from a tank. This air moves through the respiratory system with ease as the diver continues to breathe, even as the pressure around them increases with further descent. Now, what would happen if the diver reaches the ocean floor, takes a deep breath in, and then holds it as they ascend all the way to the surface? Boyle's law states that as the air pressure decreases with this ascent, the gaseous nitrogen will increase in volume—the air will expand! It is this terrifying prospect which all divers are trained to avoid by never holding their breath while ascending, and which Robert Boyle could surely have predicted.

# THE KINETIC THEORY OF GASES

Though the existence of atoms had been predicted by such prominent scientists as Isaac Newton and Robert Boyle, by the eighteenth century the notion that all matter was composed of tiny, indestructible objects was still very much unconfirmed. As physicists and chemists explored the properties of gases more intently, they began to realize that the key to understanding these gases lay in understanding something much smaller and more fundamental: the motion of atoms.

**Fluid Dynamics**

Dutch-Swiss mathematician Daniel Bernoulli (1700–1782) rose to prominence within the field of mathematics and physics in the years after Newton's death. Bernoulli's work proved vitally important to the advancement of the math involved in studying the dynamics of fluids and gases.

In 1738 Bernoulli published one of the first fully statistical explanations of gases. Bernoulli formulated what would eventually be called the kinetic theory of gases. The young mathematician discovered that he could better explain the behavior of gases by viewing them as if they were composed not of one continuous fluid, which was terribly difficult to understand mathematically, but rather of large numbers of individual microscopic particles, constantly in motion and bumping into one another, bouncing here and there in completely random directions. In other words, he calculated the behavior of gases as if they were made of atoms.

To Bernoulli, the pressure of a gas was a measurement of the number its particles and of the speed at which they strike a given surface, such as the walls of a pressurized container. Although, within any sample of gas, there are far too many of these particles to count or to calculate individually, it is possible to gauge their collective behavior statistically by using mathematical ideas well understood at that time.

This statistical analysis gave physicists a new set of methods by which to explore the nature of particle motion. This in turn led to the formulation of the laws of thermodynamics, which govern the nature of important concepts such as heat and entropy (see pp. 70–71). All of this bolstered the argument for atomic theory.

**Turning Heat into Motion**

After the work of Bernoulli, the kinetic theory of gases—especially in respect to heat—would achieve real prominence in the work of nineteenth-century physicists like Sadi Carnot and Rudolf Clausius. Carnot, a French physicist, became known as the

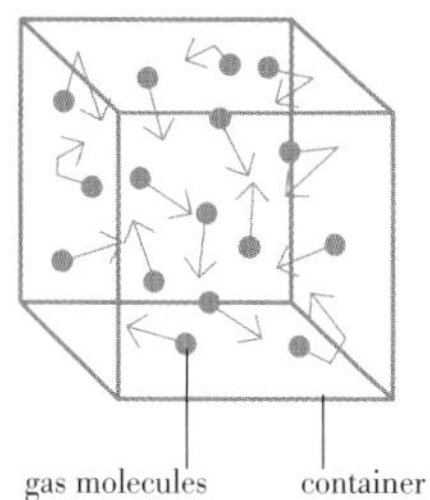


• The pressure of a gas within a container is caused by the motion of the individual particles (molecules) within the gas.

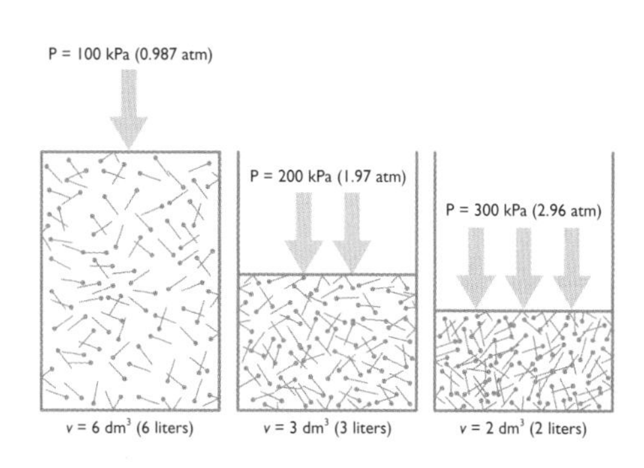


• As gas is compressed into an area of smaller volume, so the pressure increases.

father of thermodynamics after he explained the physical processes that occurred within the newly invented steam engine. In analyzing this new technological breakthrough, Carnot was able to come to his own conclusions about how the atomic nature of gases led to actual mechanical work.

The way a steam engine works is remarkably simple: water is heated to boiling point, which converts it to steam. When water transforms into steam its volume increases dramatically as a result of the increased motion of its molecules—precisely as Bernoulli's kinetic theory predicted. When this steam in a sealed chamber cannot expand, the pressure will increase, and this pressure is used to push a piston, thus transferring heat energy (or, the kinetic motion of the gas molecules) into pure mechanical energy, or motion.

Modern combustion engines work very much the same way, though rather than utilizing simple heat energy from the boiling of water, they utilize chemical energy, making use of the combustion of certain chemicals (such as gasoline or diesel fuels) to create the pressure necessary to drive pistons.

## KINETIC THEORY IN ACTION

Have you ever heated up a pot of water so much that the lid actually starts to shake and rattle under the pressure of the steam? If so, then you've experienced the power of the kinetic theory of gases. When the pressure of the steam builds up to this extent, the motion of the pot's lid is caused by the motion of individual water particles within the steam itself. The hotter they get, the faster they move, and the more steam is created, the more pressure all of these jostling atoms will place on the pot lid. No, we cannot see atoms, but we can readily see their effects.

# THE LAWS OF THERMODYNAMICS

What is heat? What is energy? The kinetic theory of gases, which described the phenomena of heat and pressure in terms of tiny microscopic particles all jiggling around in a substance, led not only to further evidence for the existence of atoms, but to a greater understanding of these ideas as well. The resulting laws of thermodynamics, which were formulated in the nineteenth century, defined precisely how heat and energy behave and how we can use them to our advantage.

**The First Law**

*The increase in the internal energy of a system is equal to the amount of energy added by heating the system minus the amount lost as a result of the work being done by the system.*

This rather cumbersome law is also known as the "law of conservation of energy." Essentially, it is saying that the amount of energy within the universe will never change. All energy must go somewhere and the total energy of any situation can always be accounted for.

Looking carefully at any situation reveals the truth of this law. Imagine, for example, a car rolling down a hill and crashing into a tree. The car possesses a lot of kinetic energy (motion) as it rolls down the hill, but all of this energy seemingly disappears as soon as it strikes the tree. Where did it go?

The energy did not disappear–it merely dispersed and changed forms. The energy of motion was transformed into sound, heat, and the motion of individual particles within the vehicle that caused the structure of the car to become deformed. Many forms of energy exist and they can all be converted into one another because energy is nothing more than motion. It is the movement of particles, and as Newton's laws of motion clearly state, motion can be transferred from one object to another. So the particles in a car can easily transfer their motion to the particles in a tree but the motion will never disappear entirely.

**The Second Law**

*In a system, any process will tend to increase the total entropy of the universe.*

First off: what is entropy? Entropy is simply disorder. It is chaos. In other words, the total amount of disarray in the universe is always increasing.

In any system (whether it be the human body, a person's home, or a planet) there is a certain amount of order. Atoms are joined with other atoms

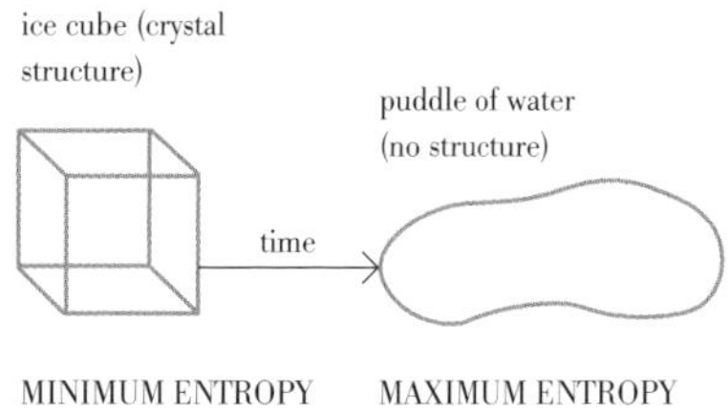


to form molecules, complex bonds create complex forms of matter, animals interact with other animals, plants grow, and so on. But what the second law of thermodynamics says is that this order is steadily decreasing.

Consider a mirror falling off a wall and breaking into hundreds of pieces. This constitutes a decrease in order, for all these nicely aligned particles within the glass have now been fractured. Try as we might, it is virtually impossible to fully restore the order to this system. In fact, in almost every situation it is far easier to go from order to disorder than the other way around. In any system, at least some energy is lost to entropy, contributing to the total chaos of the universe (for more on this law see the box at the bottom of the page).

### The Third Law

*As temperature approaches absolute zero, the entropy of a system approaches a constant minimum.*

The third law of thermodynamics was developed in the nineteenth century by William Thomson (later known as Lord Kelvin), several decades after the first two. Thomson recognized that, if heat is caused by the motion of particles, then there must be a point a which all heat is lost—that is, when the particles come to a complete stop. This point is known as absolute zero, or, on the scale that Thomson created, 0 degrees Kelvin.

#### THE FUTILE SEARCH FOR PERPETUAL MOTION

The second law of thermodynamics tells us that entropy is always increasing—that is, that energy is always being lost. Because of this fundamental principle, every machine ever built requires a supply of energy in order to keep moving. Though the search for a device in which motion continues on forever without ceasing (that is, the search for a "perpetual motion" device) has been going on for hundreds of years, it is this principle which tells us that such a device will surely never be found. In any machine there will be energy lost, whether to friction, sound or some other form.

Exercise 8

# Temperature and Thermodynamics

## THE PROBLEM:

Lord Kelvin decides to experiment with his new refrigeration device by placing several different materials into it. He first places a piece of metal cutlery (a spoon) and a piece of cardboard into the icebox for several hours. He then takes them out several hours later and feels them both, only to discover that while both items have been in the same environment for the same amount of time, the metal feels much colder than the cardboard! What is going on here and what does this tell us about the physics of heat?

## THE METHOD:

In order to solve this peculiar problem, it is important for Lord Kelvin to understand a very important principle of thermodynamics: heat conduction.

When any two objects come into contact with each other, they immediately begin moving toward "thermal equilibrium." This means that heat will be transferred between the two objects (which is not limited just to solid bodies, but anything of "substance," including gases and liquids) until thermal equilibrium has been reached.

Thermal equilibrium occurs when heat is no longer exchanged between two objects. In other words, it is when both have achieved the same temperature.

## THE SOLUTION:

Using this knowledge Lord Kelvin can actually experimentally verify whether the two objects placed into the freezer are really different in temperature or if this is merely an illusion. To do this he needs nothing more than a thermometer.

A thermometer is essentially a device which utilizes the principle of thermal equilibrium to tell us the temperature of an object. All Lord Kelvin must do is place the thermometer against each of the objects removed from the freezer, which will decrease the temperature of the mercury within the thermometer, causing its volume to decrease and making it "fall" until it reaches thermal equilibrium. As soon as the heat energy is no longer being transferred between the object and the thermometer the mercury stops falling and we can tell just how cold the object is.

What does Lord Kelvin find when he does this? He finds, in fact, that their temperatures are precisely the same! The difference in temperature he feels is due to the materials themselves—metal, because of its atomic structure, is a much more efficient conductor of heat than cardboard. When Kelvin touches the metal, heat is transferred from his body much more rapidly than is the case with the cardboard, even though they are themselves the same temperature. This makes the metal feel much colder. We can see evidence of thermal equilibrium everywhere in our world. When we drop an ice cube into a hot cup of coffee in order to cool it down the coffee quickly transfers some of its heat to the melting ice until thermal equilibrium is achieved, leaving the coffee a bit cooler.

• The Kelvin scale was named for the work of William Thomson, 1st Baron Kelvin.

Why does our hot chocolate get cold before we finish drinking it? It has been trying to achieve thermal equilibrium with the outside air. The entire universe is perpetually seeking to achieve total thermal equilibrium, and we have learned to use this to our advantage.

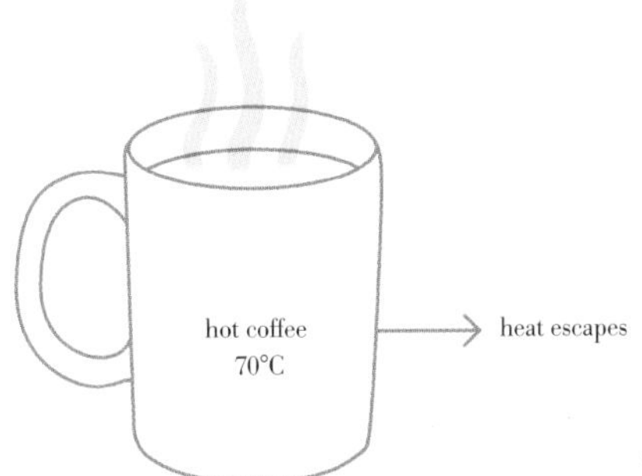


• Heat flows in the direction of decreasing temperature.

Profile

# Michael Faraday

**Michael Faraday perhaps best exemplifies the science of the nineteenth century. His work helped to usher in a new era of science and technology—his work in chemistry led to incredible advancements in our knowledge of matter and his work with electricity contributed much to modern technology. Among other things, he discovered electromagnetic induction, the dependence between electricity and magnetism, and he invented the electric motor.**

## The Bookbinder's Assistant

Born in 1791 to a poor family in London, Michael Faraday, like a great many of the men who would later become prominent scientists, demonstrated extreme curiosity from a very early age.

As a child Faraday attended school during the day, learning reading, writing and basic mathematics. At 13 years old Faraday had to find work to help the family finances and was employed running errands for a local bookbinder. After a year of this, in 1805, Faraday was taken in as an apprentice, where he spent seven years, both binding and reading the many books which fell into his hands. Reading some of the latest works in science led Faraday to an early interest in physics.

Soon Faraday began attending local scientific lectures covering such topics as electricity (see pp. 80–81) and mechanics—the latest frontiers in scientific research. In 1812 Faraday attended lectures by Humphry Davy—one of the era's most respected scientists and teachers—at the Royal Institution. The young bookbinder made careful notes of these lectures and later created a beautifully bound volume of them, which he sent to Davy. These notes so impressed Davy that when the position opened up in 1813, he hired Faraday as his assistant at the Royal Institution.

## At the Royal Institution

Faraday accompanied Davy on a scientific tour of Europe, where they met many of the most prominent European scientists of the day, such as Ampère (in France) and Volta (in Italy)—two men whose names have become synonymous with matters of electricity.

Returning to London, Faraday and Davy went back to work, with Faraday demonstrating immediately just how proficient he was at both experimenting and in giving lectures.

In 1821 Faraday married Sarah Barnard and in 1824 he was elected a fellow of the Royal Institution, though this promotion had actually been opposed by Davy (the society's previous president) in a fit of professional jealousy.

Three more decades of research followed, though Faraday would not live to see the culmination of his life's work and the rise of modern electronics. He died in 1867.

## The Work of Faraday

The focus of Faraday's most influential work centered on the concepts of electricity and magnetism. By 1820 scientists had worked to establish a relation between these two phenomena. Davy became interested and this gave Faraday the opportunity to work on the topic.

Though he was never known as a strong mathematician, Faraday was able to experimentally establish the direct relationship between electricity and magnetism. This would lead others to develop mathematical theories of light and electricity.

Faraday's research led to the discovery of electromagnetic rotation—a continuous circular motion from the circular magnetic force around a wire. Ten years later he began his great series of experiments in which he discovered electromagnetic induction. Faraday used an "induction ring" to generate electricity in a wire by using the electromagnetic field produced by a current in another wire. This was the first electric transformer. Later that year he also performed experiments whereby he produced a steady electric current by way of magneto-electric induction. He discovered that a changing magnetic field could induce an electrical current in a nearby wire—the first electric generator. By inducing motion in a magnetic field, Faraday produced an electric current and almost single-handedly ushered in the modern age of electronics.

Faraday's research wasn't wholly confined to electricity, of course. In the ten years from 1821 to 1831 Faraday's research on chemistry led to the liquefaction of chlorine and isolation of benzene among many other things.

### GALVANI'S DEAD FROGS

One of the most wildly popular demonstrations of the power of electricity was developed by the Italian physician Luigi Galvani. He would make the limbs of a dead frog twitch by applying an electric charge to the spinal cord. The resulting effect, which appeared to bring the frog back to life, amazed audiences and opened up new and exciting electrical mysteries.

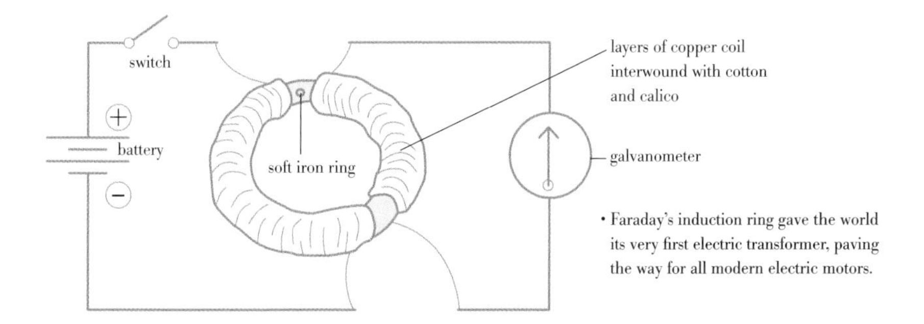


• Faraday's induction ring gave the world its very first electric transformer, paving the way for all modern electric motors.

Chapter

4

# Physics in the 1800s

The nineteenth century was a crucial period in the development of physics. During that time the world witnessed the first theories regarding electricity, the formal discovery of the atom, and the very first attempts to delve into the subatomic realm with the discovery of the electron and of radioactivity. These breakthroughs led to tremendous practical application as well as new physical theories. Much of what we currently know about physics has its roots in the work performed during this century.

Profile

# James Clerk Maxwell

**While Michael Faraday's work laid the foundations for a fully scientific theory of electromagnetism, it was James Clerk Maxwell who first developed the mathematical theory of light and electricity. He demonstrated that concepts as complex as light and electricity can be fully explained using science and mathematics.**

## Maxwell's Life

James Clerk Maxwell was born in Scotland in 1831. Maxwell's mother, Frances, was responsible for his early education, though she died from abdominal cancer when her son was only eight. Maxwell's father, John, took over his education, hiring a tutor to teach his son—though this does not seem to have been particularly successful as the tutor was fired shortly afterwards.

In 1841 his father sent him to stay with his aunt while he attended the prestigious Edinburgh Academy. It seems that Maxwell did not fit in well at school, and was mocked by his classmates because of his thick accent and relatively humble background. Yet despite his personal difficulties, Maxwell demonstrated remarkable intelligence even at this early age, becoming fascinated with mathematics (especially geometry). At the age of 13 he won awards for mathematics, English, and poetry. When he was only 14 years old, Maxwell wrote his first mathematical paper, detailing a means of drawing mathematical curves using a piece of twine and explaining the properties of ellipses and curves with more than two foci.

Maxwell continued to prosper in his studies. He graduated from Trinity College, Cambridge, in 1854 with a degree in mathematics and went on to hold professorships at Marischal College in Aberdeen and King's College in London. By 1871 Maxwell was named the first Cavendish Professor of Physics at Cambridge.

It was this period, from the 1860s to the mid-1870s that Maxwell performed some of his most important work analyzing and creating a detailed mathematical description of electro-magnetism, which would become his greatest legacy in science.

In 1879, Maxwell, who had been suffering from abdominal cancer, died in Cambridge at the age of 48.

"Maxwell's equations have had a greater impact on human history than any ten presidents."

***—Carl Sagan***

## Maxwell's Science

Though Maxwell is remembered today primarily for the equations he developed to explain electromagnetism, the full range of his mathematical and scientific contributions is actually quite extensive.

Maxwell's first major contribution was in astronomy, with a study of the planet Saturn's rings. He theorized that the rings consisted of small solid particles in order to maintain stability in their orbit around the planet (which was proved to be correct). He went on to work on statistical explanations of moving particles (see pp. 88–89), eventually making essential contributions to a mathematical model of the kinetic theory of gases (alongside his contemporary, Ludwig Boltzman, which is why the resulting mathematics are known as Maxwell-Boltzman Statistics). This lent greater weight to the notion that heat is a product of the motion of particles (atoms).

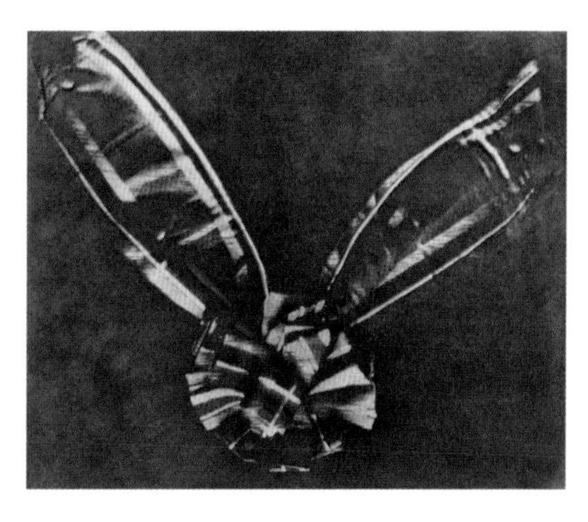

• Maxwell is credited with the first ever color photograph, produced by taking three monochromatic pictures and projecting them atop one another.

Maxwell's work was not limited to the theoretical, of course. A number of his achievements have profoundly affected our everyday lives. He is credited with having taken the first color photograph—of a tartan ribbon in 1861—by taking three different monochromatic photographs of the same object and then superimposing them on top of one another. He also performed important mathematical work in the fields of mechanics and engineering.

Maxwell's most important achievement, however, was his extension and mathematical formulation of Michael Faraday's theories of electricity and magnetism (see pp. 72–73). Conducting much of his research between 1864 and 1873, Maxwell showed that the behavior of both electric and magnetic fields (and the relationship between the two of them) could be expressed using only a few relatively simple mathematical equations. It wasn't their simplicity, however, that made them so momentous.

These equations were remarkable because they finally demonstrated mathematically that electricity and magnetism are not merely related, but they are one in the same. Maxwell showed that an oscillating electric charge produces an electromagnetic field.

Maxwell's equations are one of the great achievements of nineteenth-century physics. In fact, Maxwell's work on electricity and magnetism (known collectively as "electrodynamics") is seen by many as the most important scientific advance of that period. It has arguably had the greatest impact on the way people lived their lives in the twentieth century because it enabled electricity to power the computers, vehicles, and spacecraft of the future. All users of modern technology have Maxwell to thank.

# MAXWELL'S EQUATIONS

Richard Feynman, one of the most prominent physicists of the second half of the twentieth century, paid tribute to James Clerk Maxwell in one of his famed lectures: "From a long view of the history of mankind—seen from, say, 10,000 years from now, there can be little doubt that the most significant event of the nineteenth century will be judged as Maxwell's discovery of the laws of electrodynamics. The American Civil War will pale into provincial insignificance in comparison with this important scientific event of the same decade."

### The Electromagnetic Question

Prior to the work of Maxwell, light remained something of a mystery. In the same way, before Michael Faraday's work on electromagnetic induction, both electricity and magnetism were poorly understood. It was Maxwell who saw that all of these phenomena could be explained mathematically, and by doing so he made the rather astounding observation that they could all be articulated using the same equations.

In Maxwell's theory, light is nothing more than a repeated oscillation between electricity and magnetism. Faraday had shown that electricity and magnetism are one and the same—light is a product of the interaction between these two things. Inside a ray of light one may imagine a complex interaction between electricity and magnetism: electricity is produced, which creates a bit of magnetism, which creates a bit of electricity, and so on. According to Maxwell, the result of this interaction between electricity and magnetism is not just the single ray of light but an entire electromagnetic field.

### The Equations

In 1865, Maxwell offered his theory in the form of eight equations, each of which explained a small piece of the electromagnetic puzzle. Today these equations have been simplified to four (without changing any of their results), which look like this:

**Gauss' Law for Electricity**

$\nabla \cdot E = \rho / \varepsilon_0$

Named after German mathematician Carl Friedrich Gauss, this first equation defines the change of an electric field (denoted as "$\nabla$"). This equation is used to quantify the behavior of an electrical field in terms of its charge density in a given point of space at a given time. The change of an electric field is equal to its charge density ($\rho$) divided by a universal constant, $\varepsilon_0$ which is equal to

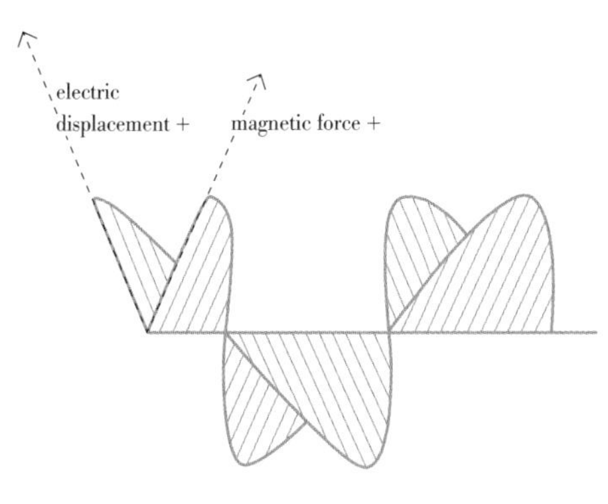


• Maxwell's equations see light as a "leapfrogging" interaction between waves of electricity and magnetism.

about 8.85 x $10^{-12}$ coulombs (a coulomb is a unit of electrical charge).

**Gauss' Law for Magnetism**

$\nabla \cdot B = 0$

This one is fairly simple: It simply states that the overall change ($\nabla$.) in a magnetic field $B$ is always zero. In other words, magnetism is always flowing from one pole to another (such as from the positive end to the negative end of a magnet), but because the magnetism is always moving in loops, nothing is ever gained or lost. The total change is zero.

**Faraday's Law of Induction**

$\nabla \times E = -\partial B/\partial t$

This is a mathematical description of Faraday's invention of the electric motor. It says that the "curl" of an electric field is dependent on the rate of change of a magnetic field. This is what Michael Faraday noticed when he was able to induce an electric charge in a loop of wire by moving a magnetic field in and out of it—he was inducing "curliness" in the electrical field and thus causing electricity to move in the circular wire. This is the equation which determines the behavior of an electric motor or generator! Very important.

**Ampère's Law**

$\nabla \times H = J + \partial D/\partial t$

The final equation is the opposite of Faraday's law of induction. It says that a changing electric field will affect the "curl" of a magnetic field. Faraday said that a changing magnetic field will induce a "curly" electric field, while Ampère's law says that a changing electric field will induce a "curly" magnetic field. This is exactly how Maxwell realized he could explain the phenomenon of light, with these two equations working together! A bit of electricity induces curl in a magnetic field, which induces an electric field, which induces a magnetic field, and so on.

Just like that, Maxwell had explained light ("electromagnetism"), in all its many varied forms. He had explained visible light, radio waves, gamma rays, and X-rays (which had not yet been discovered), and he had even given us a means of calculating just how fast this entire process takes place, thus allowing a mathematical method of deducing the speed of light!

It's no wonder that Maxwell's equations are viewed as such a revolutionary achievement.

# THE BASICS OF ELECTRICITY

Of all of the scientific advancements in the nineteenth century—whether in atomic theory, chemistry, astronomy, or radioactivity—perhaps the subject which has most directly shaped the world in which we all live is the study of electricity. A large percentage of the world's population uses electricity daily—it sparked revolutions in manufacturing, in communication, and succeeded in simply making all of our lives a bit easier. But what exactly is it?

## The Basics

Though the existence of electricity has been known since ancient times (the word itself comes from the Greek word *elektron*, meaning "amber," because of that material's ability to create static electricity), it is only relatively recently that it has been understood. Today we know that electricity is simply the flow of electrons through matter.

Electrons are negatively charged particles that surround the positively charged atomic nucleus. The negative charge of an atom's electrons is usually equal to the positive charge of the nucleus. When the balancing force between protons and electrons is upset by an outside force, an atom may gain or lose an electron. When electrons are "lost" from an atom, the free movement of these electrons constitutes an electric current.

Some types of atoms, such as metals, allow electrons to flow easily (known as conductors); some do not allow electron flow at all (insulators); some, such as silicon, allow only limited flow ("semiconductors"). As scientists began to understand these properties ("electrical resistance") of various elements, the electrical revolution was sparked.

## Harnessing Electricity to Do Work

Today we can do a great many things with electricity—we can directly power machines which perform difficult and laborious tasks for us; we can create moving images on screens and their accompanying sounds; we can send and receive signals almost instantaneously anywhere in the world;

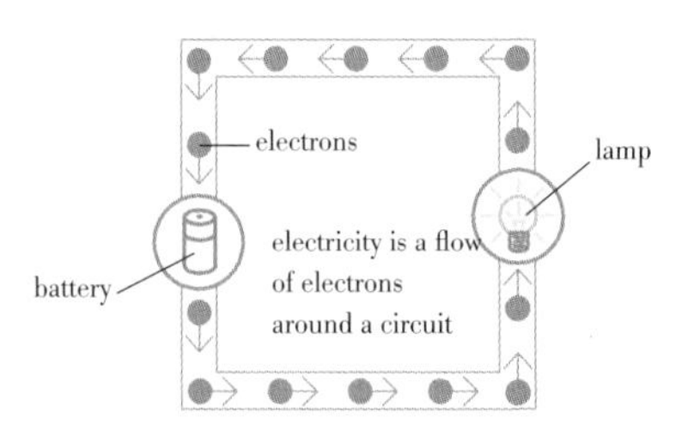


• Electrical conduction is dependent on a material's electron configuration—some materials, such as metals, are strong conductors, while others resist the flow of electricity.

we can even rub our heads with balloons and make our hair stand on end.

By the middle of the eighteenth century it had already been recognized that objects could be pushed or pulled by magnetism. Electricity had been passed through wires several miles in length (showing that a message could, theoretically, be sent using electrical impulses, which led to the telegraph and all modern forms of communication) and patients were already being shocked with electricity in the hope that it might hold some therapeutic value.

It was when Faraday first demonstrated the principle of electrical induction in the middle of the nineteenth century, however, that it was first realized that electricity could actually be made to do real work. He had shown that electricity could be created by way of the rotation of a magnet—a principle that, before long, was being used on larger and larger scales.

Today's electric utility power stations operate on precisely the principles discovered by Faraday more than a century and a half ago, using either a wind turbine, gas-powered engine, water wheel, or any other machine to drive an electric generator that converts mechanical or chemical energy into electricity.

At the opposite end of the spectrum, where magnetism is produced by way of electricity, we have produced powerful electromagnets which can lift large metallic objects (such as in scrap yards) or, as is common today, to direct the flow of particles, as in a particle accelerator.

### ELECTRON VOLTS

Particle physicists have defined a new, more precise unit of energy that is derived from the electrical charge of the electron. This unit, called the "electron volt," is equivalent to the energy transfer of an electron passing through one volt.

The electron volt (eV) is commonly used to quantify the energy produced by collisions within particle accelerators. Because an electron volt is so tiny, these collisions often take place at energies equivalent to billions or even trillions of electron volts, and so are typically used in magnitudes of gigaelectron volts (GeV, or one billion eV) or teraelectron volts (TeV, or one trillion eV).

## From Electricity to Electronics

Today, the use of electricity varies wildly, from the traditional (light and heat, for example) to the much more novel, such as the complex circuitry found in today's electronic devices. But no matter how complicated electronics become, we are still considering nothing more than the flow of electrons within atoms, making use of the electrical properties of elements and transforming them into transistors, resistors, and diodes which can regulate the flow of power down to the individual electron. We've certainly come a long way, but the principles remain the same.

# Volts, Watts, Amps, and Ohms

## THE PROBLEM:

An inquisitive inventor is working on a new appliance which will use a tremendous electrical source. While attempting to calculate the precise amount of energy required to run his appliance, he finds himself getting bogged down by the number of different terms used to describe the capacity of a simple electrical circuit. Finally, he stops to consider the ways of describing electrical phenomena. What, he asks, are watts, volts, amps, and ohms and why are they important? What do these terms have to do with the energy usage of his new appliance and the wiring running through his laboratory?

## THE METHOD:

The first step is to note the differences between the various units involved in measuring electricity. These can, thankfully, be reduced to four basic units: volts, amps, watts, and ohms.

"Volts" are units which measure the "pressure" of electrical current. Voltage standards differ between countries (the U.S., for example, uses 120 volt systems, while the UK uses 230 volts).

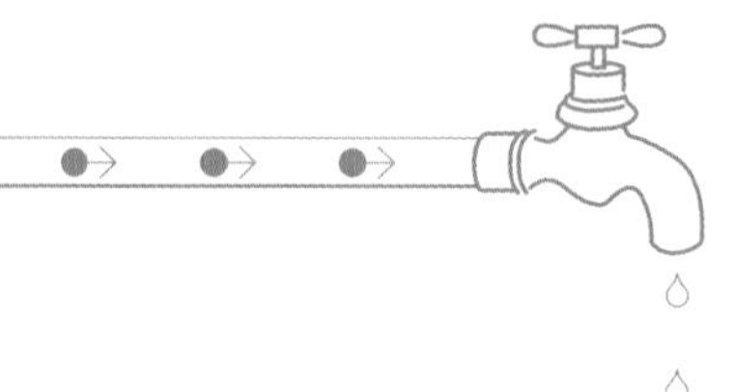

• The flow of water through pipes and spigots provides a simple analogy for the flow of electricity through wires and circuits.

• Opening the valve is like increasing the voltage of a circuit, thus increasing the electrical current.

"Watts" are units by which to measure electric power—that is, the ability of the electricity to actually do work. Watts are very important to the inventor, as they determine just how much electricity his new device will use and how much more his customers will owe on their electric bill. The average home in the United States is wired to allow around 1,800 watts, and each electronic device, appliance and light bulb in the home will use a certain number of these watts (a standard incandescent light bulb will use between 60 and 100 watts, while a refrigerator might use closer to 700 watts).

"Amps" (short for amperes) are used to measure the "flow" of the electric current (how many electrons are actually passing through the circuit).

"Ohms" are units which measure the resistance of a material to the flow of an electric current.

Most importantly, however, is to recognize that none of these units can be considered on their own, for they are all related directly to one another by way of very simple mathematical equations:

- The relationship between power (watts), voltage (volts), and current (amps) is Watts = Volts x Amps.
- The relationship between voltage (in volts) to current (in amps) and resistance (in ohms) is known as Ohm's law: Voltage = Amps x Ohms.
- The wattage of any system is therefore dependent on both its voltage and amperage, while the voltage of a system is dependent on its amperage and resistance (in ohms).

## THE SOLUTION:

The inventor finds that the relationship between the various units used to measure electricity can perhaps best be understood using a basic plumbing analogy. He imagines a hose connected to a house's water supply. When the valve is opened a bit, water flows easily through the tube and out the other end—this is just like electrical current flowing through the wires in a home.

When the valve is opened more, the pressure of water increases and more comes out. This is exactly what happens when the voltage of an electrical system is increased. Increasing the voltage makes more current (in amps) flow through the wire.

But what happens when you increase the diameter of the hose? The amount of water flowing through will remain the same, but the pressure will decrease. In electricity, this is the same as increasing the electrical resistance (ohms)—it will not change the amperage but will decrease the voltage.

Using this relatively simple picture is just the first step in understanding the physics of how electricity works within the home and within electronic devices. All electrical systems are merely tools for regulating and managing volts, watts, amps, and ohms. Using this knowledge, the inventor will be able to make the electrical system in his new invention as efficient as possible, so as to decrease the wattage and save his customers money on their electric bill.

# THE QUESTIONS OF LIGHT

From the very beginning, the difficult subject of light was shrouded in mystery. Is it a substance? Is it a force? Does it move or does it simply exist? Does it travel at a finite speed? If so, just how fast is it? Though the questions were asked early on, the answers would not be found until well into the Renaissance period.

**Defining the Speed of Light**

In 1638, Galileo Galilei attempted to determine the speed of light by measuring the delay as it passed between lanterns on two hills. The experiment was a failure, however, and led Galileo only to the conclusion that either light traveled infinitely fast, or that it was simply too fast to be measured in this experiment.

Only 40 years later, a Danish astronomer named Ole Rømer made a monumental discovery as he was studying the moons of Jupiter, specifically in regard to the innermost moon, Io, which had tremendous implications for the study of light.

In 1676, Rømer recognized that Io's orbit around Jupiter was not always consistent from his perspective on Earth. He correctly deduced that this might be caused by the delay from the light traveling from Jupiter to the Earth. Rømer was able to offer the first ever calculation of the speed of light to roughly 220,000 kilometers per second—certainly closer than anyone before had come to the presently agreed value.

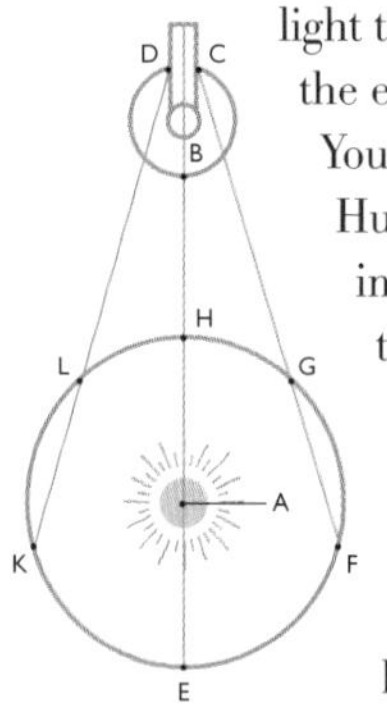

• Ole Rømer observed the orbit of Io around Jupiter (B–D), relative to Earth's orbit around the Sun (E–H, K, and L).

**Refining the Concept of Light**

By the eighteenth century, a finite value for the speed of light had been accepted and the question itself had changed, turning instead to the very substance of light itself.

Isaac Newton, among others, believed that light consisted of particles (which he called corpuscles), while the Dutch physicist Christiaan Huygens believed that light traveled as waves. A century later, the experimental work of Thomas Young would favor Huygens and show that light did indeed move as a wave, and this theory was soon adopted by the majority of physicists.

The details of light's substance remained yet unclear until James Clerk Maxwell published his monumental theory of electromagnetism in 1864, demonstrating that light waves were nothing more than electromagnetic waves. These equations could also be used to mathematically derive the speed at which these forces interact, which just so happened to precisely match the measured speed of light.

### The Universal Constant

Calculations of the speed of light were further refined through the nineteenth century, eventually bringing us to the currently accepted value of 299,792.458 kilometers per second, but we also came to understand some very peculiar things about this speed.

Through the experiments of Armand Fizeau in 1851, and Albert Michelson and Edward Morley in 1887 (see "The Michelson and Morley Experiment"), a rather peculiar fact became clear: The speed of light was not a quantity that could be measured relative to one's own speed, but was actually a universal constant. This means that the speed of light is always the same, no matter what speed a person is traveling relative to the light. If one person is stationary while another is moving 10,000 kilometers per hour, the measured speed of light will remain the same for both individuals.

While it may seem counterintuitive, it was this realization that would eventually lead Albert Einstein (see pp. 106–107) to form his theory of special relativity in 1905 (wherein the speed of light was included in a great many of equations as the constant *c*), and which would herald a new era in physics.

### THE MICHELSON AND MORLEY EXPERIMENT

In 1887, Albert Michelson and Edward Morley designed one of the most important experiments in the history of physics. At that point it was still believed that the universe was filled by an invisible substance, called the "ether" (first theorized by Aristotle). Michelson and Morley attempted to detect the presence of ether by seeing if light traveled more slowly when moving against the ether than when traveling with it. They discovered that not only did the ether not have an effect on the speed of light (leading many to believe that it didn't exist in the first place), but that the measured speed of light did not change at all relative to the motion of the Earth! This experiment eventually helped lead to Einstein's special theory of relativity (see pp. 108–109).

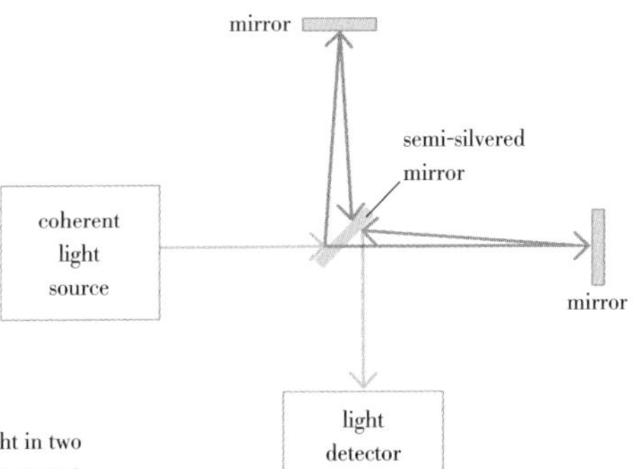


• The Michelson/Morley experiment measured the speed of light in two directions, using three mirrors, one of which was partially transparent.

# UNCUTTABLE THINGS—ATOMS

Though the earliest atomic theory originated more than 2,500 years ago with the Greek philosopher Democritus, this idea was effectively killed by the "elemental" theory of Empedocles, which remained dominant well into the "modern" era of science. In the eighteenth and nineteenth centuries, however, atomic theory staged a comeback.

### The Atom Reborn

Though the existence of atoms had been speculated about for centuries, it was the English chemist John Dalton (1766–1844) who finally presented a fully workable, pragmatic view of atomic theory. By taking extensive measurements of the pressures and masses of many different gases he was led to believe that their properties might be dependent on the tiny little particles (atoms) from which they were formed. Dalton went on to theorize that different atoms of differing weights could account for all of the various known elements.

Consequently, Dalton was the first physicist to attempt to create a table of atomic weights (a rough one in 1803, then a better one in 1805)—a task which would be expanded upon to near perfection by Russian chemist Dmitri Mendeleyev more than 60 years later, resulting in a periodic table very similar to the one used today.

John Dalton

John Dalton also developed such tenets of atomic theory as the fact that all atoms of a given element are identical in every way; no two elements share any common atoms; and atoms can bind together to form compounds.

### Dalton's Imperfect Theory

There was one nearly fatal flaw in John Dalton's logic regarding atomic theory, however. In his list of atomic "rules," Dalton stated that these particles must, by definition, be indivisible—as the name itself means "uncuttable" in Greek. He stated that there could be nothing smaller than an atom, despite the fact that there was not yet (nor would there ever be in the future) any real evidence that this was the case. In his mind, he had found the most basic building blocks of the universe.

Dalton turned out to be quite wrong about this but nonetheless made great headway with the still rather primitive scientific tools at his disposal. John Dalton could surely not have known what was in store for atomic theory. While he died relatively confident that his atoms did, in fact, exist, he could not possibly have foreseen the discovery of their complex structure, their many parts, or the forces that bind them all together. He could never have guessed that the theory he played such an important role in forming would have led to the Large Hadron Collider—a multi-billion dollar particle accelerator designed to probe the depths of these tiny little objects.

## The Size of Atoms

It is certainly not difficult to understand why it took so long for atomic theory to catch on. After all, there were no practical methods available to scientists prior to the nineteenth century to discover these tiny little things, apart from perhaps mathematics, and even then, in the work of Dalton and others, such experiments had to be exceptionally clever. To this very day, atoms remain, by definition, invisible to science.

For something to be seen by the human eye, it has simply to be larger than the wavelength of light (for things are seen by the light reflecting off of them). Any smaller than that and the thing is invisible. The shortest wavelength of visible light is just under 400 nm (that is, nanometers, or billionths of a meter). In comparison, the widest of all atoms sits comfortably somewhere around 500 pm (trillionths of a meter). In other words, it would take more than a thousand of the largest possible atoms laying side by side for them to be even possibly seen by the most powerful optical microscope in existence. And yet, these are the objects out of which all matter is made.

Fortunately for us, modern technology has brought about ways for us to "see" atoms (or at the very least, to see their effects), with such devices as particle accelerators and highly advanced microscopes (called "microscopes"). Today we are certain that atoms do, in fact, exist, but we also know that they are not nearly as simple as John Dalton implied. They are complex machines with moving parts and intricate interactions between a variety of forces which simultaneously stick them together with incredible force and attempt to tear them apart. Atoms, in short, are one of nature's most fascinating works of art.

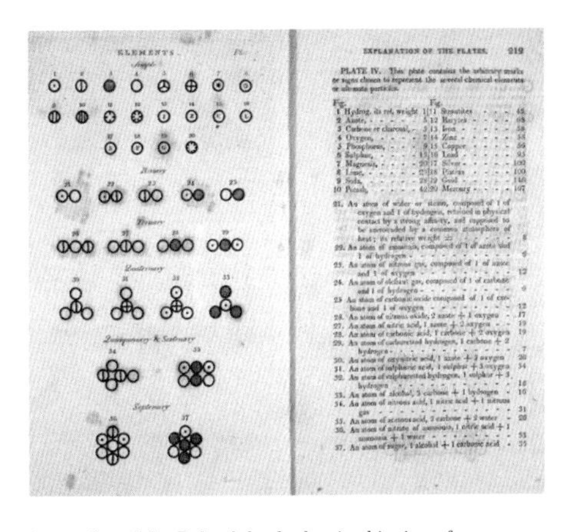
EXPLANATION OF THE PLATES. 219

PLATE IV.

• A page from John Dalton's book, showing his view of several sizes of atoms, consisting of various configurations of the smallest atom, hydrogen.

# BROWNIAN MOTION

Even after the atom was discovered, the rather farfetched concept of a universe made up of an uncountable number of tiny particles (atoms), all bound together by some unknown force to create all matter, remained rather difficult for many reputable scientists to believe. While John Dalton's atomic theory was clever, and certainly very helpful to chemists, it had yet to be experimentally verified. Until, that is, the remarkable work of the Scottish botanist Robert Brown.

### Experimental Evidence of Atoms

In his 1827 observations, Brown noticed that pollen grains placed into a clear liquid and viewed through a microscope tended to behave somewhat strangely, moving around almost as if they were alive, jumping and skipping about randomly with no visible impetus or discernible pattern.

This effect is what today is known as Brownian Motion, and was soon discovered to be caused not by the pollen grains being alive (as was a common explanation of the time), but by the grains being pushed around by the molecules within the water itself—in other words, they were being "jostled" about by molecules.

Brownian motion did much to verify the existence of tiny particles even in a seemingly continuous fluid such as water. The significance of this wouldn't be known until nearly 80 years later, when Albert Einstein took the concept of Brownian motion to the next level.

### Einstein's Contribution

One of Einstein's four papers published during his "miracle" year of 1905 was entitled "A New Measurement of Molecular Dimensions & On the Motion of Small Particles Suspended in a Stationary Liquid." While the title of Einstein's work was not exactly memorable, the work's contents are undoubtedly interesting.

By examining Brownian motion, Einstein was able to calculate the number of water molecules per square inch to a surprising degree of accuracy, as well as to provide statistical and mathematical formulas for the motion that was evidenced.

His theory was based on the assumption that, as particles move about in a liquid, a pressure is exerted on them by even smaller particles in every direction. Normally, there are roughly the same number of atoms on each side of the pollen grain, all pushing and bumping against each other randomly, so such movement should tend to cancel itself out most of the time. Seeing as it truly is a random process, however, it is not infrequent that the pollen grain is pushed a bit more in one direction, so it moves that way, then later it is pushed in a different direction and so moves another way.

## The Random Walk

To understand a bit more about Brownian motion, it helps to consider a large balloon being thrown into the audience of a concert. The balloon is bounced from one part of the stadium to the other, hit in random directions by the excited fans. Though at several points perhaps it seems to be moving in one particular direction, it soon turns and goes the opposite way, but in the end it really doesn't go anywhere, it just jostles about. A giant looking down upon this scene might not be able to see the tiny spectators in the audience—he would just see a little dot (the balloon) moving randomly.

Einstein determined that while such movement is completely random and unpredictable, it also obeys certain laws of probability, which Einstein was able to determine using a mathematical formula which became known as the random walk. Atoms had become "visible" for the first time!

### THE DRUNKARD'S WALK

Einstein's random walk formulation demonstrated that Brownian motion was indeed governed by the completely random motions of atoms within a liquid—but the applications of this random motion extend far beyond atomic physics. In fact, this same formula has been applied to a wide variety of subjects, such as:

• The formula for the "random walk" results in the same random, chaotic motion exhibited by tiny particles pushed around by atoms within a liquid.

**Genetics**—Defining the changes in gene pools over time.

**Gambling**—The probability of winning and losing are akin to a random walk.

**Economics**—Used to model the fluctuation of share prices.

**Intoxication**—The random walk can be used to model the seemingly random movements of a drunk individual attempting to find their way home from a bar. For this reason the model is often called the Drunkard's Walk.

# THE DISCOVERY OF THE ELECTRON

The atom was discovered at the beginning of the nineteenth century. Although it would take well over a century before the scientific community at large was finally convinced of the existence of atoms, physicists had already begun to probe deeper into matter—only to discover even smaller things. The first of these subatomic particles was the electron.

**Thomson's Experiment**

Credit for the discovery of the electron goes to British physicist J. J. Thomson—a man who is known not just as a great physicist (he won a Nobel Prize in 1906), but also as a phenomenal teacher, having had seven of his students (and his own son) go on to win Nobel prizes themselves.

Thomson's discovery of the electron came in 1897 while he was experimenting with a cathode ray tube (an electronic device where a beam of particles are sent between a positive and a negative terminal within a vacuum tube, creating a strikingly peculiar light (see "Cathode Rays" opposite).

• J. J. Thomson was both a great experimental physicist and a great teacher of physics.

It was uncertain at the time exactly what sort of substance cathode rays may have consisted of, but Thomson's experiments determined that whatever it was, it possessed an electric charge (he was able to bend the ray using a magnetic field and then measure the direction the ray bent so that he could determine both how much mass was involved and how it was charged). It became clear to Thomson that these rays were actually made up of tiny particles—particles which had a mass much smaller even than an atom (which could be determined by the amount of bending in a given magnetic field).

So, for the first time, it had been determined that something smaller than an atom existed—in the form of tiny negatively charged particles, though it wasn't immediately certain that these particles were actually part of the atom. They were at first thought to be a separate entity altogether.

## Reaction to the Electron

At first, many scientists rejected Thomson's assertion that he had discovered something even smaller than the atom. Eventually the evidence convinced them of the electron. One experiment after another was performed which proved conclusively the existence of electrons, although no one knew exactly what they looked like, what their function was, or why they existed at all. Many of the same questions are still being asked today.

J. J. Thomson couldn't have known what he was getting into at the time—he certainly could never have realized the importance of his discovery. Electrons, it is now known, are fundamental pieces of the atomic structure. They provide the necessary balance of charge in order to "neutralize" the atom and it is they who form all the bonds between atoms. In other words, without electrons, molecules would never form and atoms would stay separate.

Electrons are not merely the source of electrical current (though they play that highly important role as well)—they are the reason that matter exists in the first place.

Thomson's experiment was indeed crucial, and jump-started an entire wave of subatomic discoveries, including the next step in understanding the atom: the discovery of the atomic nucleus.

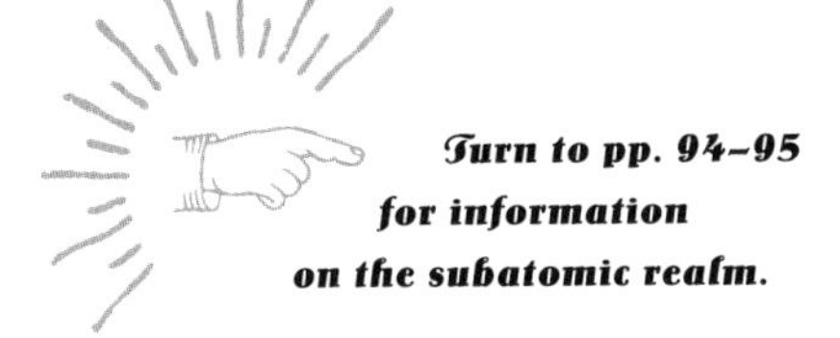

*Turn to pp. 94–95 for information on the subatomic realm.*

### CATHODE RAYS

Most people are more familiar with cathode rays than they think. In fact, before the advent of modern "flat screen" televisions, cathode rays were the driving force behind the creation of a picture on a television set. In these "older" televisions, rays of electrons (cathode rays) are emitted from an electrically-charged piece of metal and directed by magnets to strike one of thousands of multi-colored phosphorescent dots littering the screen, causing that dot to light up. Repeat this process thousands of times every second, and you have the makings of a moving picture. And that's it! The very same tool by which J. J. Thomson discovered the electron still finds ample use today.

• A cathode ray uses positively and negatively charged electrodes at either end of a vacuum-sealed tube and sends electrons between them.

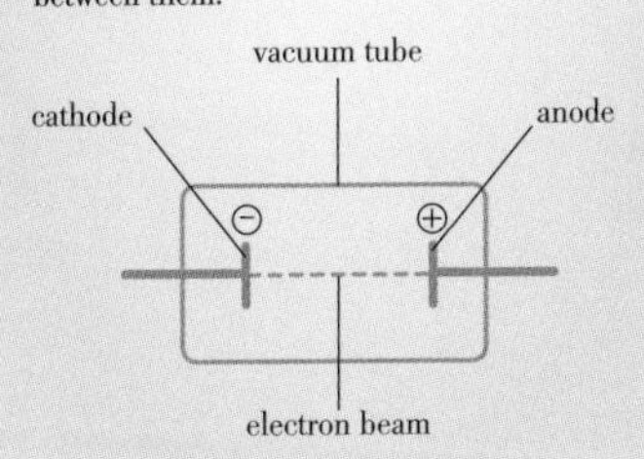

**Profile**

# Ernest Rutherford

**Known as the father of nuclear physics, Ernest Rutherford was responsible for an astonishing number of discoveries in the field of particle physics (a field which, consequently, he helped to invent), including the discovery of the atomic nucleus, proton, and the three forms of radioactive decay, the prediction of the neutron, and the creation of the modern atomic model.**

## The British Kiwi

Though a New Zealander by birth, Ernest Rutherford (1871–1937) was born to parents who had originally emigrated from Europe. His mother, Martha, was English, and his father, James, was Scottish. Rutherford achieved great success within the educational system of his homeland, earning his Bachelor of Science degree at the University of Canterbury (in Christchurch, New Zealand) before moving to Cambridge, England, in 1895 to continue his studies at the famed Cavendish Laboratory, becoming a student of J. J. Thomson (see pp. 90–91).

It was in England that Rutherford began to carry out his initial investigations into the new field of radioactivity (only recently discovered by Henri Becquerel, see pp. 98–99). His experiments with uranium and thorium led him to coin the terms alpha particles and beta particles for those two forms of radioactivity already discovered. A third, which would become known as gamma radiation, was not quite as evident in the beginning (due to its lack of electrical charge) and would not be discovered until the work of Paul Villard in 1900.

"All science is either physics or stamp collecting."

***—Ernest Rutherford***

Soon after he made these discoveries, Rutherford moved to Canada in 1898 to take up a position at McGill University in Montreal. He remained in Canada for nearly ten years before moving back to England in 1907 to work at the University of Manchester. During these years Rutherford won the 1908 Nobel Prize in Chemistry "for his investigations into the disintegration of the elements, and the chemistry of radioactive substances." In 1914 he was knighted by the British monarch. Still more honors and declarations would follow as Rutherford continued to perform work of the highest caliber.

Finally in 1919 he accepted one of the most prominent positions in all of physics—successor to J. J. Thomson as Cavendish Professor of Physics at Cambridge. He remained in Cambridge with his wife, Mary, and daughter Eileen until his death in 1937.

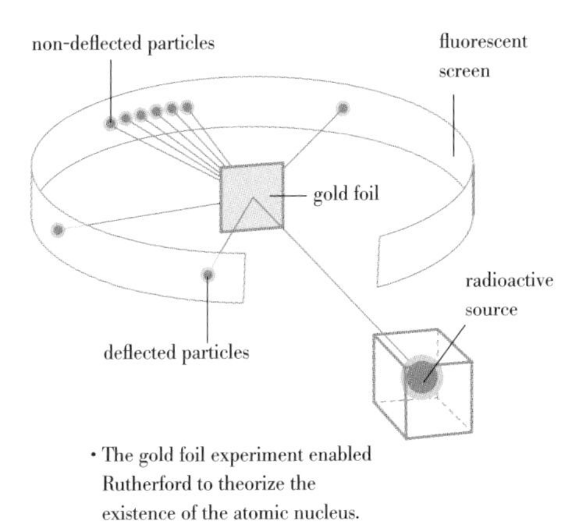


• The gold foil experiment enabled Rutherford to theorize the existence of the atomic nucleus.

## RUTHERFORD'S LEGACY

Ernest Rutherford's contributions to both physics and chemistry continue to be felt today, and he was one of the most highly decorated scientists of his era.

- He is one of the few individuals (alongside Albert Einstein, Enrico Fermi and Marie Curie) to have an element named after him—the radioactive Rutherfordium.
- He has had a crater on the Moon named for him.
- Numerous institutions, including Rutherford College in New Zealand and Rutherford College in the UK have been named for him.
- Several institutions have named buildings in his honor.
- Streets in the US, UK, and New Zealand have been named for him.
- His image appears on the $100 bill in New Zealand.
- He has been the subject of a major play.

And the list goes on . . .

## The Gold Foil Revolution

Though winning the Nobel Prize in 1908 might have marked the high point of many scientific careers, for Rutherford this was no excuse to slack off in his research. In fact, the very next year saw Rutherford and his students, Hans Geiger and Ernest Marsden, undertaking an experiment (known as the "gold foil" experiment) which led to his prediction of the atomic nucleus. Not long after this, Rutherford predicted that the atomic nucleus was made up of positively charged particles (protons), which he helped to discover, and even predicted that there might also be a neutrally charged counterpart to protons. The neutron was later discovered by one of Rutherford's former students, James Chadwick, in 1932.

Not only was Rutherford primarily responsible for the creation of the first atomic model (the aptly named "solar system" model) but he was also partially responsible for the improvement on this model developed by one of the first "quantum physicists," Niels Bohr, who moved to England from Denmark in order to study with the great Rutherford. By adding quantum mechanics to Rutherford's model, these two great minds—Rutherford one of the last great nineteenth-century thinkers and Bohr one of the first great minds of the twentieth—developed the Rutherford-Bohr atomic model, which helped to jump-start quantum mechanics, thus bringing science into an entirely new era.

# THE SUBATOMIC REALM

The discovery of the electron by J. J. Thomson in 1897 opened up an exciting new world of possibilities within the atom, and before long it became clear that there might be other undiscovered particles within the atom. Soon enough physicists had begun to develop theories of what an atom actually looks like and new experiments heralded the discovery of still more pieces to the atomic puzzle.

**Plum Pudding and Orbiting Electrons**

The first viable atomic model, credited to J. J. Thomson himself after his electron discovery, was the so-called "plum pudding" model, which paints an amusing mental picture of an atom as a "pudding" of positive matter, with little "plums" of negative electrons scattered throughout.

In 1909, Ernest Rutherford presided over a famous experiment which put the plum pudding model to the test. The experiment consisted of alpha particles (a type of naturally-occurring radiation consisting of positively charged helium atoms) being fired into a very thin sheet of gold foil. The trajectories of these particles after passing through the foil were then detected. If the plum pudding model of the atom was correct, then it was assumed that many of the particles passing through would have had their courses slightly altered by the charge within the atoms.

The result of the experiment differed remarkably from expectation. While most of the alpha particles passed right through the foil as if it wasn't even there, a small fraction of them were diverted at very dramatic angles as if striking a solid surface and bouncing off!

After analyzing the data, Rutherford realized that the plum pudding model could not account for this discrepancy, and in 1911 he proposed a new and improved atomic model—the Rutherford model. In his version of the atomic structure, he posited that the negative electrons orbited around a tiny, positively charged, and incredibly dense central "nucleus," like planets orbiting a tiny sun. While it is known today that there are certainly limitations to how accurately this model explains what is

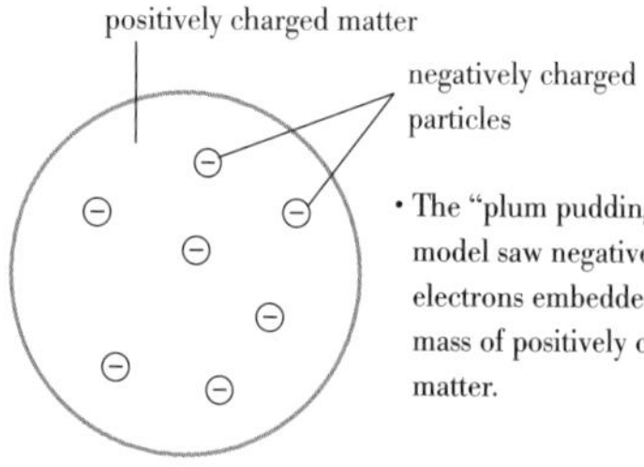

positively charged matter

negatively charged particles

• The "plum pudding" atomic model saw negatively charged electrons embedded within a mass of positively charged matter.

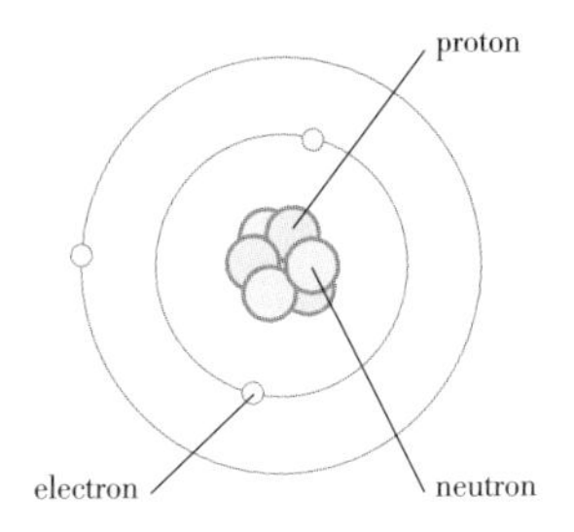


• The "solar system" atomic model was a significant step toward today's model, with electrons orbiting a hard, dense nucleus made of both protons and neutrons.

really happening within an atom, it has proved tremendously useful in explaining certain fundamental features of the atomic structure and its behavior.

## What Is in the Nucleus?

The first question Rutherford addressed was what precisely the nucleus was made of. He knew that whatever it was had to be positively charged and very dense, but this left a lot still to be explained. The nucleus was the next great frontier in atomic theory.

In 1918, Rutherford performed another experiment, bombarding nitrogen gas with alpha particles, leading to a surge of hydrogen. He correctly deduced that the hydrogen atoms must have come from within the nitrogen atoms themselves, which meant that there was something within all of these atoms which was divisible, so that a lighter element could be "removed" from a heavier element. This particle was the proton, which by itself constitutes the nucleus of a single hydrogen atom. The proton in the nucleus was recognized to be the positive antithesis to the electron, thus creating a neutral atom.

At this point, the atom seems nearly complete, but of course today we know that there remains a very important piece missing.

## The Neutron

This final piece was not discovered until 1932, though Rutherford (who else?) had been contemplating its possible existence as early as 1920. The existence of a neutral particle with roughly the same mass as the proton would fix some disparity found between an atom's number and its mass (for the atomic number is defined by the number of protons, where the atomic mass, which measures the mass of the nucleus, was generally higher).

In 1932, English physicist James Chadwick (a student of Rutherford's) discovered the neutron after performing a series of tests on a new type of radiation which had been baffling physicists for years and which had previously been mistaken for gamma radiation.

Chadwick bombarded a sample of Beryllium with alpha particles, causing it to emit this mysterious radiation. The radiation would strike a proton-rich surface and some of the protons would be discharged which could then be detected using a Geiger counter (a device that measures radiation). At this point in the experiment, Chadwick did a little detective work: He knew that the radiation was neutral after exposing it to a magnetic field, and it had to be somewhat heavy in order to discharge something as heavy as protons. Certainly it was heavier than gamma radiation (which has no mass at all). The radiation was made of neutrons.

For his achievement of completing the basic atomic model, Chadwick received the Nobel Prize in 1935.

Profile

# Marie Curie

**Marie Curie was a rarity—a female scientist in a field dominated by males. Despite this disadvantage, her work was brilliant. So good in fact, that Curie has the rare distinction of winning not just one but two Nobel Prizes, in both physics and in chemistry. She, along with her husband Pierre, helped to invent the field of atomic physics—a field whose mysteries are still being discovered today.**

## Early Life

Born Maria Skłodowska in Warsaw, Poland, on November 7, 1867, the woman who would become Marie Curie was the daughter of a school teacher. From her father she received basic schooling, including simple skills in science. In 1891 Marie left Poland to study in Paris at the Sorbonne, where she excelled in both physics and mathematics. Within just a few years of moving to France she had met Pierre Curie, Professor in the School of Physics, and was married.

Marie and Pierre Curie formed one of the few husband-and-wife teams of scientists to have obtained prominence within the world of physics (another such Nobel Prize-winning team would consist of the Curies' daughter Irène and her husband Frédéric Joliot).

Upon graduating from the Sorbonne with outstanding scores, Curie became interested in the recent discovery of radiation by Henri Becquerel. Curie began studying uranium radiation, using new and revolutionary techniques (some devised by her husband), and soon enough Pierre had ceased his own research in order to join his wife.

By 1898 the Curies, just five years into their marriage, were already announcing the discovery of two new elements, Radium and Polonium (the latter of which Marie named after her native Poland) and Marie became the first person to use the term "radioactive."

Radioactivity became the singular focus of Curie's professional life. After the tragic and untimely death of Pierre in 1906, she went on to promote the use of radium to alleviate suffering during World War I (along with her daughter Irène). She later sought to establish a radioactivity laboratory in her native Warsaw, which she accomplished with help from the United States and a gift from President Hoover in 1929. The institute opened in 1932.

Marie Curie was held in great honor and esteem by her fellow scientists. She was a prominent voice at scientific conferences and achieved a considerable profile internationally, winning numerous awards. Curie succeeded her husband as Head of the Physics Laboratory at the Sorbonne. After its founding in 1914, she was appointed Director of the Curie Laboratory in the Radium Institute of the University of Paris. Unfortunately, the radioactivity to which she dedicated her professional life proved to be her

| H | | | | | | | | | | | | | | | | | He |
|---|---|---|---|---|---|---|---|---|---|---|---|---|---|---|---|---|---|
| Li | Be | | | | | | | | | | | B | C | N | O | F | Ne |
| Na | Mg | | | | | | | | | | | Al | Si | P | S | Cl | Ar |
| K | Ca | Sc | Ti | V | Cr | Mn | Fe | Co | Ni | Cu | Zn | Ga | Ge | As | Se | Br | Kr |
| Rb | Sr | Y | Zr | Nb | Mo | Tc | Ru | Rh | Pd | Ag | Cd | In | Sn | Sb | Te | I | Xe |
| Cs | Ba | | Hf | Ta | W | Re | Os | Ir | Pt | Au | Hg | Ti | Pb | Bi | Po | At | Rn |
| | Ra | | Rf | Db | Sq | Bh | Hs | Mt | Ds | Rg | | | | | | | |
| | | | La | Ce | Pr | Nd | Pm | Sm | Eu | Gd | Tb | Dy | Ho | Er | Tm | Yb | Lu |
| | | | Ac | Th | Pa | U | Np | Pu | Am | Cm | Bk | Cf | Es | Fm | Md | No | Lr |

• While most elements possess certain isotopes that succumb to radioactive decay, the shaded elements are known to be the most radioactively unstable, having no stable isotopes at all.

downfall. She passed away in Savoy, France on July 4, 1934, after suffering from pernicious anemia brought on by her exposure to excessive amounts of radiation.

## Curie's Legacy

The achievements of Marie Curie are numerous, and they fall mainly within the one subject which intrigued her more than any other: radioactivity. The discovery of what would later be named "radioactivity" by Henri Becquerel in 1896 became the inspiration for Curie's work, for they immediately set about deciphering the mysteries of uranium radiation found in pitchblende (a naturally occurring, uranium-rich radioactive ore). Their first great achievement was in recognizing that the radiations emitted from pitchblende were even more intense than from uranium itself, which led to the realization that there were elements present even more radioactive than uranium—this led to the discovery of highly radioactive radium and polonium.

Marie Curie went on to develop methods for the separation of radium in sufficient quantities to allow for a full study of its properties, including its therapeutic benefit for patients suffering from cancer. This ushered in one of the many practical uses for radioactivity and led to great achievements in the medical sciences.

Marie and Pierre Curie shared the Nobel Prize for Physics with Henri Becquerel in 1903 for their study of the radiation discovered by Becquerel. Eight years later (after Pierre's death), Marie Curie won a second Nobel Prize, this time in Chemistry, in recognition of her work in radioactivity. She was the first, and is now one of only four individuals, to have won the award twice.

**"Nothing in life is to be feared, it is only to be understood. Now is the time to understand more, so that we may fear less."—*Marie Curie***

# RADIOACTIVITY

Most of us will have heard of radioactivity. We know that it is dangerous and probably best to avoid. We know that it is left over from atom bombs and from nuclear power plants and that it is hugely controversial. But what of the science of radioactivity? The study of this phenomenon has been ongoing for more than a century, but we still have much to learn about this powerful and mysterious force in which atoms spontaneously begin to break apart.

## The Early Science of Radiation

One of the early pioneers of the study of radioactivity was French physicist Henri Becquerel. Like his father before him, physicist Alexandre-Edmond Becquerel, Becquerel was fascinated by the subject of atomic phosphorescence (substances that emit a glow, especially after having been energized by exposure to light), and performed tests on various phosphorescent compounds.

In 1896, Becquerel noticed an odd thing when he was experimenting on potassium uranyl sulfate (uranyl implies that the compound contains uranium). He found that the compound would "radiate" even without being exposed to the Sun, leaving traces on a photographic plate nearby, even if wrapped in an opaque material.

Whatever was coming out of the compound penetrated right through various materials like X-rays (a form of high-frequency electromagnetism discovered only a year previously). It was clear that whatever was causing this radiation had to be some heretofore undiscovered property of the chemical compound itself. Becquerel had discovered atomic radiation.

Becquerel's work was continued by Marie and Pierre Curie. Marie Curie showed that several different compounds were radioactive and that they all seemed to have the element uranium (atomic number 92) in common. She later found that thorium (atomic number 90) was also radioactive and discovered two new radioactive elements—radium and polonium. Curie's work led to the understanding that only certain forms of the heaviest elements were radioactive.

## Alpha and Beta Radiation

The next great step came from Ernest Rutherford. Rutherford, who would later be commemorated with his very own radioactive element—Rutherfordium, noticed in 1898 that radiation came in at least two different types, which he called alpha and beta, after the first letters of the Greek alphabet. What's more, Rutherford

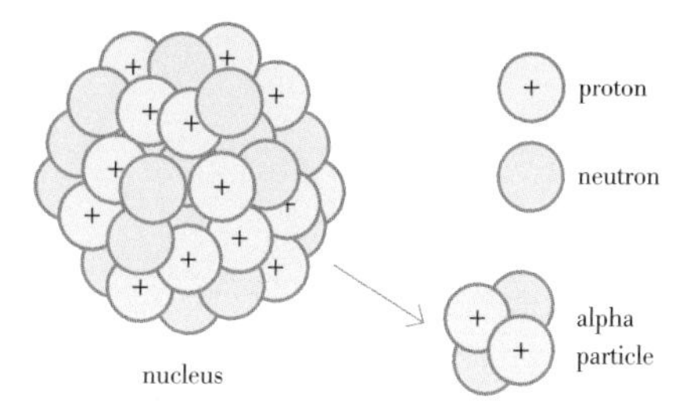


noted the different properties of these types of radiation: alpha particles, which Rutherford believed to be hydrogen nuclei (though later tests showed that they were actually helium nuclei) and beta particles, which are escaping electrons.

A third type of radiation, consisting of highly energetic electromagnetism (light) was discovered in 1900 by French physicist Paul Villard and called gamma radiation. Gamma radiation also possesses a very short wavelength which makes it very penetrable. These gamma rays are much more intense than X-rays, or even the ultraviolet light which causes the Sun's rays to burn human skin, making them very dangerous.

## Spontaneous Radioactive Decay

These first nuclear physicists noticed that elements appeared to undergo radioactive "decay" for no apparent reason. Particles of radiation seemed prone to jumping randomly out of the atomic nucleus, which made it very difficult to either measure or predict.

After careful study it became clear that only certain isotopes of certain atoms (an isotope is a form of an element containing the same number of protons but a different number of neutrons), known as radionuclides, are radioactive. Radioactivity only occurs in these isotopes due to the composition of their nuclei; the combination of protons and neutrons makes them unstable.

Although at the time it was not yet clear what held the nucleus together, we are now familiar with the strong nuclear force, which provides an incredibly strong, though highly localized, bond between the components of the nucleus in a manner analogous to the electromagnetic force, which holds the electrons in place in the outer portions of the atom. Even with this strong force operating between protons and neutrons, however, sometimes it is only by a fine thread that everything is kept relatively stable, especially in the larger atoms, where the force simply does not have enough range to keep all the particles together. Study of the strong force and another force, known as the weak nuclear force (which helps to split an atom apart), would become foundational fields of twentieth-century physics and continue to intrigue some of the greatest minds today.

• In beta decay, an electron is emitted from the atomic nucleus, turning a neutron into a proton and changing the entire chemical composition of the atom, such as from carbon to nitrogen.

$${}^{1}_{0}n \rightarrow {}^{1}_{1}p + {}^{0}_{-1}e$$

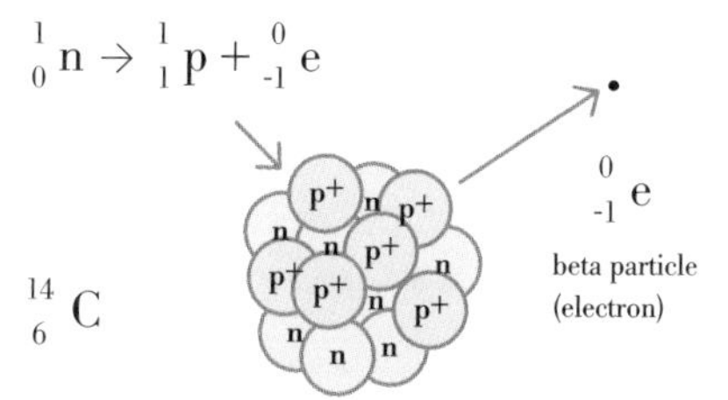

# BLACKBODY RADIATION AND THE FALL OF CLASSICAL PHYSICS

It seems that the greatest scientific advancements come not when our experiments prove our theories even more accurate, but when we find ourselves facing a situation where the experiment no longer coincides with the theory. This is precisely what happened toward the end of the nineteenth century, where a glaring problem in early particle physics demonstrated that perhaps the laws of physics were not as well-understood as most had assumed.

## What Is Blackbody Radiation?

Toward the end of the nineteenth century, many physicists thought that they had just about everything figured out. But in the 1890s, a few seemingly harmless and innocent questions arose. For example: how does a red-hot piece of metal fit into the world of classical physics?

A substance such as this red-hot metal is what physicists would call a "blackbody"—a material which absorbs without reflection all of the electromagnetic radiation (light) that hits it. While a simple piece of metal is not a perfect blackbody (those do not exist), it serves as a good example.

Despite its name, a blackbody is not always black, and this is where the problem lies. A blackbody's color, like a black metal poker stuck into a fire, is dependent on its temperature. It begins to heat up as it absorbs the energy from the fire, then it begins to glow red, then orange, and finally white as its temperature increases.

This change in color occurs because as the temperature increases, the wavelength of the electromagnetic radiation coming from it decreases (the heat absorbed by a blackbody is emitted in the form of heat radiation), thus changing into visible colors with shorter wavelengths. So far so simple.

## The Ultraviolet Catastrophe

The problem is that the laws of classical mechanics state that blackbodies which have achieved thermodynamic equilibrium (that is, when they absorb as much energy as they radiate back out) should radiate energy at every wavelength. Mathematically, this means that when a blackbody gets hot enough, the amount of radiation that is theoretically given off should begin to approach infinity. In other words, a glowing body should in theory emit a tremendous amount of radiant

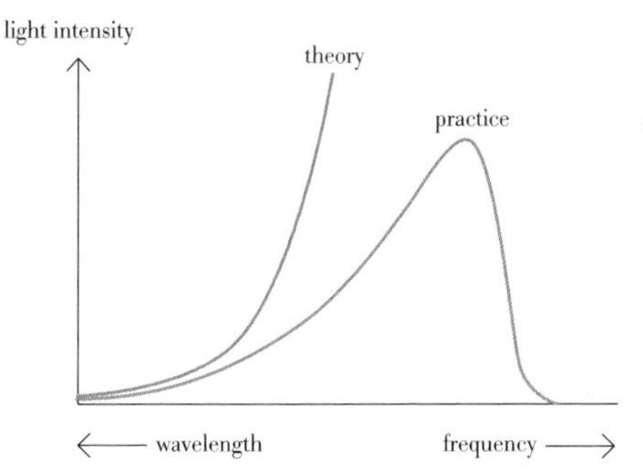


• The Ultraviolet Catastrophe occurs where practice diverges from theory; in classical physics light should be emitted from blackbodies with infinite intensity, while practical experience shows otherwise.

energy that is strong enough to burn up everything in sight. Obviously, this doesn't happen. Either something is wrong with the blackbodies themselves (not very likely), or something is wrong with classical physics (very likely).

When the actual radiation emitted from a blackbody was measured, it did not shoot toward infinity at the ultraviolet region of the electromagnetic scale (as the theories suggested), but rather reached its peak toward the middle of the visible range of the spectrum, which seemed entirely counter-intuitive. It is for this reason that the discrepancy became known as the "ultraviolet catastrophe."

## Planck's Solution

It was Max Planck who finally provided a solution to this problem by boldly searching for a mathematical formula that would seek to explain this phenomenon. What Planck came up with was a deceptively simple, yet undoubtedly revolutionary little equation (called, appropriately, Planck's law of blackbody radiation):

$$E = hv$$

E here represents the thermal radiation produced by a blackbody; $v$ (which is actually the Greek letter "nu") is the frequency of the electromagnetism being released; and a new constant, $h$, dubbed Planck's constant (which has a very tiny estimated value of somewhere very close to $6.626068 \times 10^{-34}$ J·s). And that's it—a fairly simple equation that solved one of the biggest questions facing scientists in the 1890s.

## What Does It Mean?

The true importance of this equation lies in Planck's constant itself. Though Planck himself did not understand at first what this number represented, this constant would become the cornerstone of an entirely new form of physics. Essentially, Planck's constant represents the smallest possible unit of radiation, which eventually became known as a photon, or a particle of light!

Throughout the nineteenth century it had been shown (and nearly proven) that light traveled as waves. Planck's theory implied that, within these waves, light did in fact exist as particles after all. Planck had "quantized" light. A blackbody radiates only as many quanta of light as its energy will allow, and because each individual quanta requires energy to create, most of them fall into the wavelength of visible light, where the least amount of energy is required. And just like that, with the recognition that light was actually made of particles, quantum physics was born.

Chapter

# The Birth of a New Physics

The study of physics was turned upside down at the beginning of the twentieth century by the discovery of two different, but equally revolutionary new fields of science: relativity and quantum mechanics. This chapter offers a brief introduction to each of these exciting branches of physics and demonstrates just how important they were to the continued pursuit of science.

**Profile**

# Max Planck

**December 14, 1900, may be considered the birthday of quantum physics, although even those present that day—remarkably intelligent scientists who surely understood every word of it—seemed to come away wholly unimpressed. It seems the significance of what they had witnessed was not immediately evident to them. On the day in question, Max Planck presented a paper to the German Physical Society that offered a solution to the problem of blackbody radiation. The simple formula contained within Planck's paper would transform the world of physics.**

• Max Planck introduced quantum physics to the world, albeit unwittingly, in December of 1900.

## Planck's Life

Max Planck was born in Kiel, Germany, in 1858. His father was a law professor in Kiel. Though his full given name was Karl Ernst Ludwig Marx Planck, by the age of ten his primary name, Marx, had been shortened to Max, which was the name he went by for the remainder of his life.

Planck's family moved to Munich in 1867, where he was enrolled in a local school and began to learn the basic principles of mathematics, astronomy and mechanics from a teacher named Hermann Müller. Though he was gifted in a number of areas, both academic and artistic (he was a talented musician), Planck chose to study physics at the University of Munich and later at the University of Berlin. At the former, Planck received much of his formal training in physics, and at the latter he studied under the physicist Hermann von Hemlholtz and first became interested in the field of heat theory. It was from his work in this field that his greatest and most memorable achievement would come.

In October 1878 Planck passed his qualifying exams; in February 1879 he defended his dissertation; and in 1880 he presented his habilitation thesis (the highest academic achievement, earned after obtaining a doctorate). Throughout the latter decades of the nineteenth century, Planck's studies focused on one primary area: thermodynamics and heat theory. He became enthralled by the physics of heat and in the distribution of energy spectrums of radiation, and it was this love which led to him studying the phenomena associated with blackbodies and endeavoring to solve these issues.

"Scientific discovery and scientific knowledge have been achieved only by those who have gone in pursuit of it without any practical purpose whatsoever in view."—***Max Planck***

## Planck's Greatest Achievement

When Planck finally did offer up a potential mathematical solution to the blackbody problem, its non-classical nature was rather unexpected, but at the same time not entirely groundbreaking. The physics community was quite simply unsure what to make of it; a theory which seemed to work just fine on paper, but which seemed to neglect certain physical "truths" which had previously been taken for granted. It failed to adhere to the set of classical formulas which governed the whole of physics at the time.

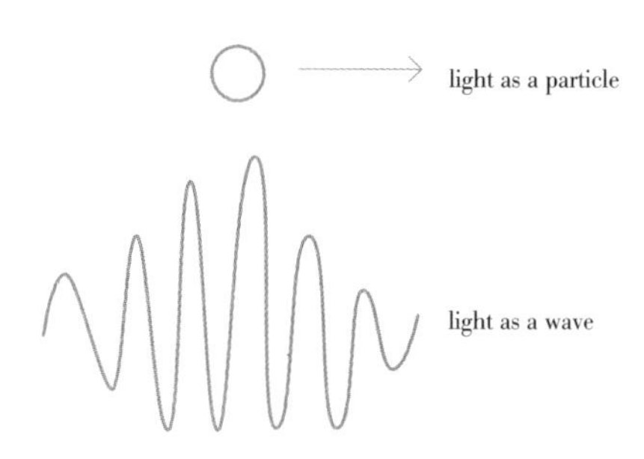


• Max Planck scarcely believed the consequences of his own research, that light could behave as both a wave and a particle.

So, what did Planck discover and why was it so difficult to digest? The equation Planck formulated described the blackbody phenomena in a purely mathematical sense, but he did not provide with it any particular physical explanation, which is what was truly sought by physicists. It was therefore not immediately clear what Planck's work signified. While figuring out how best to calculate the seemingly discontinuous nature of a very specific form of light, Planck had not yet even begun to explain what was actually happening within a blackbody that might cause it to emit radiation as it did. When he did try to explain it, he was only partially correct.

## The Incredulity of Planck

It would take five more years and the work of a young and almost completely unknown patent clerk named Albert Einstein (see pp. 106–107) for Planck's theories to gain any traction. When Einstein looked at Planck's equation, applying it to another problem that had cropped up in the previous decade, his conclusion stated what Planck's equation had only implied: light truly is quantized. It behaves as a wave but is at the same time broken up into particles, which were later referred to as photons.

Though Planck never fully accepted the quantum physics that he helped to invent, he was nevertheless awarded a Nobel Prize in Physics for his efforts in 1921 (a year before Einstein won it for his own contribution to this field) and remained a respected figurehead in physics for several decades.

Max Planck, by then in his 80s, survived World War II (though he lost a son, who was accused of plotting to assassinate Hitler). He died soon after, in 1947. He is widely considered the founder of quantum physics.

Profile

# Albert Einstein

**Albert Einstein's two great theories of relativity (which he called the "special" and the "general" theories) form one of the great branches of twentieth-century science, and help bridge the gap between classical and modern physics. They are the theories by which a virtually unknown German physicist sought to supplant Isaac Newton as history's most influential physicist.**

## Early Life

Albert Einstein was born in Ulm, Germany, in 1879 to a Jewish couple, Hermann (an electrical engineer and businessman) and Pauline Einstein. While from a young age Einstein showed signs of being gifted and possessing a remarkable mind, he did not immediately excel in his schooling. He was by nature rather introverted and was at his best when reasoning in abstraction—a fact that did not serve him well in the strict and prescriptive atmosphere of local schools such as the Luitpold Gymnasium in Munich where he attended secondary school.

As young as ten, Einstein had become heavily interested in science and began reading countless books on the subject. As a teenager he became fascinated by mathematics (specifically geometry) and had begun to ponder such abstract hypothetical questions as what it might be like to be able to travel at the speed of light. He surely could not have imagined at the time that it would be his very own work which would finally provide some real answers to this question.

## Einstein's Politics

Throughout much of his life, Albert Einstein was nearly as active in politics as in science. In order to stay out of the armed services he renounced his German citizenship at the age of 17, quitting school and moving with his family to Italy. After enrolling at the Federal Institute of Technology in Zurich, he became a citizen of Switzerland. After moving back to Germany, where he taught for several years at the University of Berlin, Einstein moved to the United States in 1933, partly in response to the rise of fascism in Europe. Einstein had long been a champion of peace,

"The most incomprehensible thing about the world is that it is comprehensible."

***—Albert Einstein***

though in the midst of World War II he famously wrote a letter to President Roosevelt urging him to move forward in the building of a nuclear weapon, for fear that the Germans might do so first. This was the extent of Einstein's involvement with nuclear weapons, but even this was enough to cause him considerable guilt after they had been used against Japan.

Einstein lived in Princeton, New Jersey, (where he taught at the Institute for Advanced Studies) until his death in 1955, just a few years after turning down an offer to become the president of the new nation of Israel.

## Scientific Achievements

Einstein's greatest scientific achievements were in physics, and included his two theories of relativity. But these were not his sole achievements. His prodigious mind also led to outstanding work in quantum mechanics (for which he won the 1921 Nobel Prize), atomic theory, and cosmology.

When he wrote the first of five groundbreaking papers in 1905, which detailed the theory that would become known as special relativity, the young Einstein was working as a patent officer in Switzerland. The year 1905 is commonly referred to as Einstein's *annus mirabilis*—his "wonderful year." Over the course of 12 months, Einstein—totally unknown in the scientific world up to that point—would publish a total of five papers. Two of these papers were on the subject of atomic theory and served to once and for all demonstrate the existence of atoms and provide means of measuring their size. Another was one of the first important papers on quantum theory for which the aforementioned Nobel Prize would later be awarded. Two more established the foundations of the special theory of relativity. Any one of these papers on their own would have been revolutionary. The five of them taken together truly do seem miraculous. It took some time for Einstein's fame to truly spread as a result of this early work, but once his achievements were picked up by the international press he became a household name.

Einstein wasn't finished, however. He quickly continued his work. In quantum mechanics he embarked on a five-decade-long quest to truly understand what this new science was trying to say, which led to intense conflicts with many of the greatest scientists of that era. In the midst of this he expanded his theory of relativity (which dealt with constant motion in a vacuum) to create a "general" version of the theory (published in 1916), which completely rewrote the book on gravity.

During the final decades of his life, Einstein's work was mostly limited to the debate over quantum mechanics and attempts to use the general theory of relativity to understand the size, shape, and behavior of the universe itself. It was a monumentally difficult task, but Einstein hoped that it would lead him to a "theory of everything," which would explain all that there was to know in physics. Einstein never achieved this grand goal, of course, and even now, more than half a century later, this remains one of the great unrealized ambitions of physics.

# SPECIAL RELATIVITY

Albert Einstein tore up the rule book when he published his theory of special relativity. This young and unknown man suddenly told the world that everything they knew about light, motion, speed, and even time were false. Though it would be years before his magnificent theory finally began to fully sink into the minds of physicists, special relativity would forever change physics and our perceptions of reality.

## The Foundations of Relativity

The special theory of relativity is built upon two foundations:

### Galilean Relativity

Galileo (see pp. 42–43) had offered a simple principle 300 years earlier: there can be no "preferred" frame of reference. In other words, the laws of physics within a moving vehicle (such as a boat, train, or spacecraft) are no different than the laws of physics on solid ground. Throw a ball inside a moving train and it will behave precisely the same as a ball thrown on a stationary train, with no scientific means of telling the difference. This is, in essence, the law of relativity.

### The Constant Speed of Light

Nineteenth-century physicists demonstrated that the speed of light is constant and unchanging, no matter what speed an observer is moving. What Einstein discovered was this: it is only because of the constant nature of the speed of light that Galilean relativity holds true. The laws of physics hold true in any reference frame, whether it is moving or stationary, for even the speed of light will not change relative to one's motion.

## Galilean Relativity and the Train Analogy

Einstein often used analogies involving trains to explain his theory; there are two reasons for this. Firstly, they move (generally) in a straight path, stuck to rails; secondly, once in motion, they generally maintain a consistent speed.

One such analogy goes like this: inside a moving train, a ball is dropped to the floor. To an observer who is on board the train, the ball would be seen to travel straight downward. To an observer outside the train, however (providing they could see through the train car's walls), the ball would appear to fall along a curved path in the direction the train is moving.

The question is which of these two viewpoints (the observer on the train or the observer on land) is correct (or as Einstein said, which is "preferred")? Is the ball falling in a straight line or is it

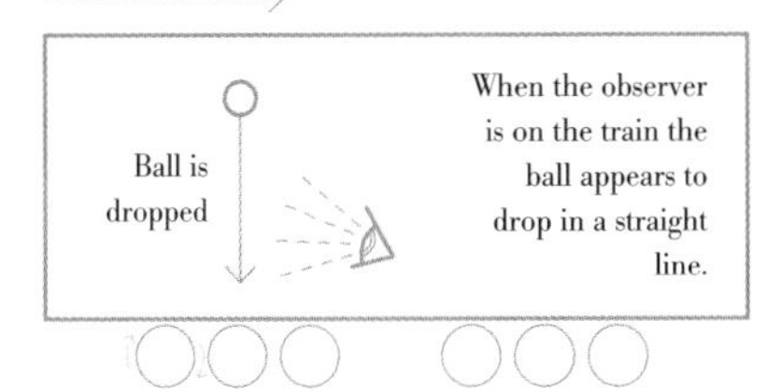


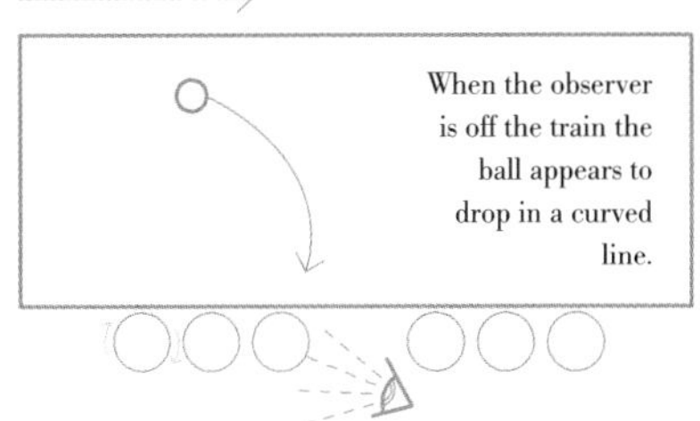


moving in a curve, as it would be relative to the Earth? In the parlance of relativity, these two different perspectives are actually equal and are referred to as reference frames or inertial frames. In the theory of relativity, all reference frames are equal, though their relative motion alters the way they are seen by outside observers.

This idea of the equality of inertial frames is the key to unlocking the secrets and mysteries inherent in Einstein's own theory of special relativity which would come almost 300 years later.

## Defining the Speed of Light as a Constant

Now, we must add to the concept of relativity the idea of the constant speed of light. Imagine that you and two of your friends are racing through space using jetpacks. You are all traveling at different speeds, one at 30 km/h, one at 300 km/h, and one at 3,000 km/h.

Consider that as the three of you are racing at your various speeds, a ray of light should suddenly come zooming past you, in the same direction, at the speed of light. If all three of you were somehow able to measure the speed of this light in relation to your own speeds, what do you suppose you would find? According to traditional methods, you would surely expect that in order to find the speed of light relative to yourself, you would simply take the speed of light ($c$), and subtract from it your own speed, right? Not so, if the speed of light is a constant. The truth is that all three of you would continue to measure the speed of this light to be exactly the same: 300,000 km/s.

What Einstein's theory does is to combine the "basic" theory of relativity with this constant speed of light. It says that Galileo's theory of relativity does not merely relate to motion, but to the speed of light itself, and if light is a constant, everything else is relative, including even perceptions of time and space. In special relativity time can slow down and speed up (see the following pages), objects expand and contract, and nothing is what it seems. Everything is relative and dependent on motion.

• To the external observer, the movement of the train creates the impression that the ball is dropping in a curved line.

# Time and Space Dilation

**THE PROBLEM:**

Albert Einstein is designing an ultra-powerful rocket that has the capability to reach unbelievable speeds. Before testing its ability to traverse the solar system he would like to determine the relativistic effects all of these high speeds will have on himself and the crew. For example, what does his own theory of relativity say will happen to a rocket traveling at 100,000 kilometers per second? How about traveling at the very speed of light itself (300,000 km/s)?

**THE METHOD:**

One of the fundamental features of Einstein's theory of special relativity is the set of predictions it makes regarding moving objects. Based solely on the premise that any moving object must take into account the fact that light will always be measured as traveling at a constant speed, some truly bizarre consequences must be considered before traveling at extremely high speeds.

Fortunately, Einstein provided an incredibly simple equation to determine what kinds of effects speed will have on his spaceship:

$$S = \sqrt{(1 - v^2/c^2)}$$

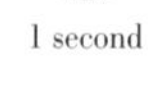

1 second

From the perspective of a fast moving spacecraft, a trip may seem to take only one second . . .

1.6 seconds

. . . while from a swiftly moving object the same trip might take up to 1.6 seconds . . .

2 seconds

. . . and from a slow moving object the same trip might take up to 2 seconds.

The $S$ in this equation is often called the shrinking factor. This is the degree by which one can measure the relative shrinkage of both time and space at various speeds. In terms of time: as the value for S rises, time speeds up (relative to Einstein's friends back on Earth, who are nearly stationary); as S decreases, time slows down.

In terms of space: as the value for S rises, the "apparent" length of Einstein's ship increases (that is, for a stationary person, his ship will appear longer); as S decreases, his ship will decrease in size.

This equation is fairly simple to use. Einstein just needs to enter in his velocity ($v$) and solve for S, which gives him his shrinking factor ($c$ stands for the speed of light–300,000 km/s).

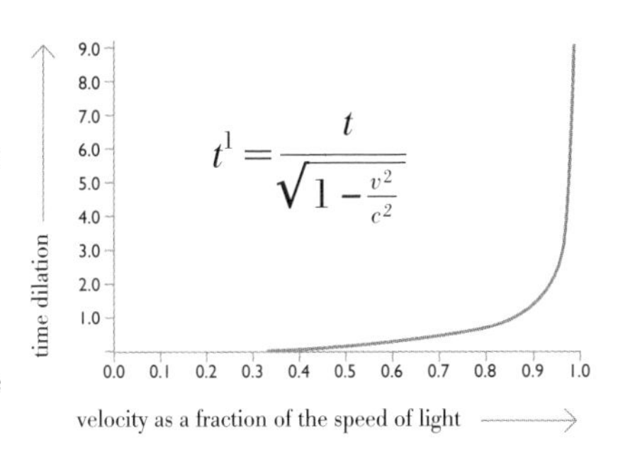


• As this graph shows, the degree of time dilation increases very slowly until an object approaches the speed of light, at which point time really begins to slow down.

**THE SOLUTION:**

So, what would happen if Einstein's rocket traveled at 100,000 kilometers per second?

$$S = \sqrt{(1 - 100{,}000^2/300{,}000^2)}$$
$$S = 0.9428$$

This means that time will be moving at about 94% of its original speed and will only have slowed down by about 6%. At the same time, the rocket's apparent length to an outside observer will have contracted by the same amount. After a full year of traveling at this speed he will have only lost about 21 total days compared to his friends on Earth.

How about at 250,000 kilometers per second?

$$S = \sqrt{(1 - 250{,}000^2/300{,}000^2)}$$
$$S = 0.55$$

Now both time and length have been cut nearly in half! For every second Einstein spends on his rocket almost two seconds will pass on Earth.

As the velocity approaches very close to the speed of light, Einstein will notice that the shrinking factor drops very quickly (290,000 km/s = 0.256; 299,999 km/s = 0.0025).

But what if Einstein traveled the speed of light? He plugs in the numbers and ends up with a problem: S = 0. He's just shrunk down to absolute nothing and time has stopped. He doesn't exist, and it seems that he will be forced to stay that way for eternity. Not exactly a pretty picture–but fortunately, it doesn't appear possible, either, so he doesn't really have to worry. This is one of the primary reasons Einstein was forced to declare that it is impossible to ever travel at the speed of light.

# Thought Exercise: The Twin Paradox

## THE PROBLEM:

Two twin brothers go their separate ways in life. One of them is adventurous, the other not so much. The adventurous twin becomes an astronaut while his conservative brother decides to spend his adult life firmly planted on Earth. Off the astronaut twin goes on his mission in an extremely fast rocket.

The first twin spends a long while in space, traveling very quickly (nearly the speed of light) to the far reaches of space, then turns around and flies back to Earth just as quickly.

What will this twin find when he arrives back on Earth and reunites with his twin brother? How does Einstein's special theory of relativity affect those attempting near-light-speed space travel?

## THE METHOD:

Special Relativity asserts that under conditions where one is traveling very quickly (such as the astronaut twin), time will actually slow down relative to someone moving slowly (such as the twin on Earth). Because of this, the twins are actually able to calculate the passage of time relative to one another and to predict the degree to which they will observe time differently.

Sure enough, when the astronaut twin arrives back on Earth, he finds that while he has only aged a few years, his brother is now an extremely old man. Only a few years have passed in the spaceship while several decades have passed on Earth!

While this seems to make sense in the mathematics of special relativity, it also leads to a seemingly severe paradox: if special relativity claims that all inertial frames are equally valid—that there can be no "preferred" frame of reference—then why is the twin in the spaceship said to be "moving" while the other is "stationary." From the perspective of the spaceship, wouldn't it seem equally plausible that Earth itself is moving while the ship is motionless?

From the perspective of the Astronaut, then, it might appear that the clock on the ship is actually running faster—exactly the opposite of the result predicted by Einstein and others. How do we reconcile this?

## THE SOLUTION:

The most important explanation for the twin paradox came from French physicist Paul Langevin in 1911 (who, consequently, was the first to develop this particular form of the thought experiment). This explanation revolves around the idea of acceleration. The fact that the twin in space has undergone acceleration (which includes both speeding up and slowing down) means that the astronaut twin is no longer in an "inertial frame." He is no longer considered equal to his brother because the motion of Earth is constant and not accelerating.

As a result of this acceleration, the twin in space does indeed possess motion relative to the non-accelerating twin on Earth, and thus is affected by the time dilation of special relativity. It is the idea of acceleration which causes the symmetry breaking between a relativistic reference frame and a non-relativistic inertial frame.

In the end, then, this story is not so much a paradox within special relativity but is rather a confirmation, as it helps to exemplify some of the fundamental features of the theory using a very accessible thought experiment.

# The Relativistic Addition of Velocities

## THE PROBLEM:

Two rocket ships are flying toward each other in the middle of space, each traveling at the terrifying velocity of 50,000 kilometers per second relative to Earth. The crews of both ships are terrified (of course) as the ships draw nearer to crashing into each other, though one crew member has the presence of mind to measure just how quickly the other ship is approaching. From this perspective, at what speed is the other ship closing in?

## THE METHOD:

What is needed to solve this problem is known as the "law of the addition of velocities." This law, prior to 1905, was incredibly simple: when two objects are moving toward each other, their velocities can be added using the formula:

$$V_1 + V_2 = V_3$$

Here, $V_1$ is the velocity of the first vehicle; $V_2$ is the velocity of the second vehicle; and $V_3$ is their combined velocity. So, if two cars pass each other on a highway, each traveling 100 km/h, they will each observe the other coming toward them at 200 km/h relative to their own speeds (100 +100 = 200). That's simple enough, and if this were truly the case, then the problem above would be shockingly easy: 50,000 km/s + 50,000 km/s = 100,000 km/s. So, under classical physics, the two ships should be approaching each other at 100,000 km/s–not a very safe speed.

But this was before special relativity. As part of Einstein's assertion that perceptions of both time and space are altered depending on an object's speed, he recognized that this fact certainly also alters the law of addition of velocities. He thus came up with a new and improved formula which will allow us to solve this problem in light of relativity:

$$V_3 = (V_1 + V_2)/(1 + V_1V_2/c^2)$$

As before, $V_1$ is the velocity of the first vehicle, $V_2$ is the velocity of the second vehicle, and $V_3$ is their combined velocity (*c*, as always, represents the speed of light–300,000 km/s).

## THE SOLUTION:

To solve the problem, we simply have to plug in our known values to the above equation:

$$V_3 = (50{,}000 + 50{,}000)/(1 + 50{,}000 \times 50{,}000/c^2)$$

$$V_3 = 100{,}000/1.027777778$$

$$V_3 = 97{,}297.2973$$

So, instead of the two ships approaching each other at a combined speed of 100,000 km/s, they will only perceive each other approaching at 97,297 km/s. Even at these high speeds, then, there seems to be only a slight change from the expected answer.

This is why it was not until the twentieth century that anyone had even a thought that this might be the case. At any speed traveled by humans there is very little discrepancy between the equation and real-world measurements.

For example, two cars driving toward each other at 60 km/h will, according to relativity, be approaching each other at a combined speed of 119.999995 km/h, which is immeasurably close to what would normally be assumed. So have no fear, this new law will have very little effect until we somehow achieve the ability to travel at speeds much closer to the speed of light. At that point, however, things will surely start to get very strange.

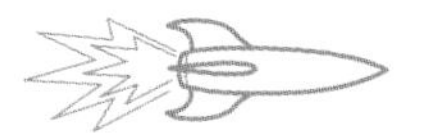

the rockets are each traveling 50,000 km/s

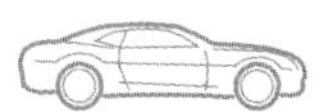
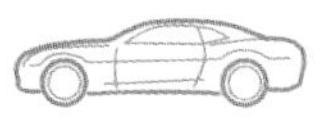

the cars are each traveling at 60 km/h

• At these relatively modest speeds, there is very little variation from the expected combined velocities.

# $E = MC^2$: THE EQUIVALENCE OF MASS AND ENERGY

It is perhaps the most famous equation in physics, though when Albert Einstein first developed his theory of the equivalence of mass and energy, it was little more than a footnote to his theory of special relativity, appearing in the shortest of the four brilliant papers he would publish in 1905, entitled "Does the Inertia of a Body Depend Upon its Energy Content?" Over the space of just a couple of pages, Einstein laid out the blueprint for the formula which would pave the way for nuclear power, high energy particle physics, and nearly everything we understand about matter and energy.

### What Are Mass and Energy?

In physics, mass is a measurement of an object's resistance to movement. The mass of an object determines how hard it is to move (this is called inertia) and its behavior within a gravitational field. Weight is a measurement of gravity acting upon an object's mass.

Energy, on the other hand, seems on the surface like an entirely different animal. Where mass is something real and tangible, energy is more of an idea. It is often defined as "the ability to do work," and while this may leave things a bit vague, it really is rather accurate.

So, when the title of Einstein's paper asks the question, "Does the Inertia (Mass) of a Body Depend Upon its Energy Content," he is asking, essentially, if there is a connection to be made between mass and energy.

### Equations for Mass and Energy

How are mass and energy connected to one another? First, they both have been known for some time to obey conservation principles, meaning that they can be neither created nor destroyed. While they can be converted into different forms (mass can turn from solid to liquid to gas and can be cut up or turned to dust, while kinetic energy can be transferred to potential, sound, or heat energy), there will always remain the same amount in our universe. They are also related, Einstein found, in another, much more fundamental way.

Einstein began, it is said, by looking at the equation for finding an object's kinetic energy:

$$E = \frac{1}{2}mv^2$$

This equation, which had been around for some time, clearly shows that there is some relationship between mass and energy, and that their relationship is defined by the velocity of an object.

Einstein, having a particularly clever year, was able to take this equation, combine it with other known equations, do a little bit of math, and come up with the following (which is starting to look more like the familiar equation, but not quite):

$$E = mc^2/\sqrt{(1 - v^2/c^2)}$$

This, in essence, is the equation Einstein published in his paper. The equation most people are familiar with, $E = mc^2$, can be obtained by assuming that the "object" in question has a speed (velocity) of zero.

$$E = mc^2/\sqrt{(1-0)}$$

$$E = mc^2$$

## Same Thing, Different Packaging

This equation shows that mass and energy are not just similar—they are the same thing, but in different forms, with the ability to be converted directly into one another. Mass can be turned into energy, and energy can be turned into mass. Furthermore, this equation shows us that a tiny bit of mass can be turned into a lot of energy (the equivalent of the amount of mass times the speed of light squared), while on the flip-side a lot of energy can only be turned into a little bit of mass.

The equation truly was a revolutionary one, though no one understood the extent of this until it was realized that it would be possible, using radioactive elements, to actually turn regular matter into pure, intense energy. The result was the atomic bomb, and later atomic energy. But perhaps even more interesting is the notion that all matter—that includes you, me, a rock, a chair, everything—is composed of pure, intense, energy! We are like living, breathing nuclear reactors!

It is Einstein's equation which helped scientists predict just what would happen when that first nuclear explosion was triggered, and it is what tells particle physicists what will happen when they smash two beams of particles together in an accelerator. In essence, it is still $E = mc^2$ that drives much of experimental and theoretical physics to this very day. Its fame, therefore, is well-substantiated.

• In a nuclear chain reaction, the atomic nucleus is split, releasing energy and more neutrons, which split other atoms, releasing yet more energy and neutrons.

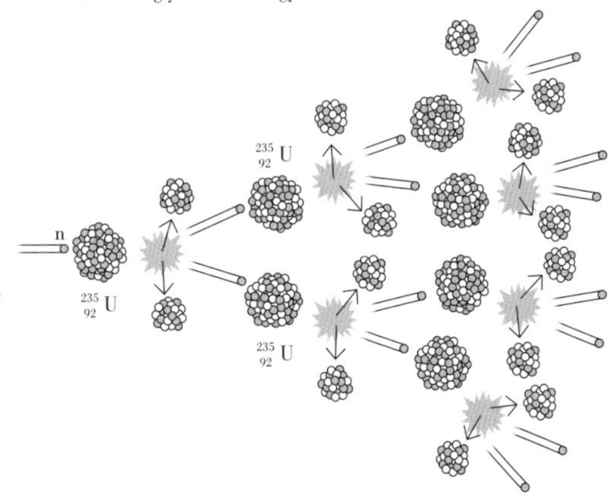

# GENERAL RELATIVITY **Part 1**

Einstein's theory of special relativity was just that—special. It was special because it was concerned only with certain very specific situations, where objects travel in perfectly straight paths and at constant speeds. For a full decade following his work in 1905, Einstein sought to expand his theory to include not just these special circumstances but all objects. The result was his theory of general relativity, which gave the world an entirely new definition of gravity.

## Einstein's Happiest Thought

In 1907, Einstein had a thought which he called "the happiest thought of my life." This idea, known as the principle of equivalence, was the result of Einstein imagining that a person accelerating through space in a ship would feel the force of the acceleration pushing against them, and to them this feeling would be no different than the feeling of being pulled down to the ground by gravity on Earth. In other words, it is impossible to tell the difference between motion and gravity.

Over the next ten years Einstein would expand upon this principle to determine that gravity must be caused, in one way or another, by acceleration—as if we are constantly "falling" into the surface of the Earth!

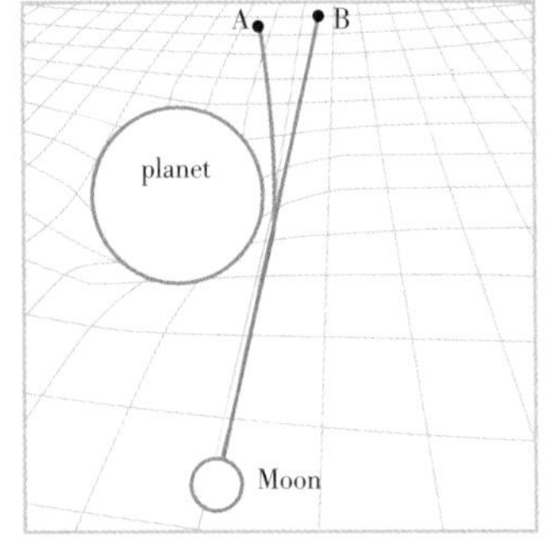


• A planet in space creates gravity by curving space-time around it, causing the path of its moon to bend, the first step in explaining orbits.

## Warped Space

How did Einstein end up solving the problem of gravity? He learned an entirely new form of mathematics, called non-Euclidean geometry, and formulated a theory in which neither space nor time is flat. Rather, in Einstein's theory, both space and time are warped by the presence of mass, which causes them to curve.

A good way to picture how gravity works within general relativity is to picture a bowling ball on a trampoline. As the ball sits on the flexible surface, it creates an indentation. In general relativity, the same thing happens when a massive object (like Earth, the Moon, or the Sun) sits in the "fabric" of space-time. Space and time must both "curve" around it, and it creates an indentation in the very dimensions of the universe. Just as with a bowling ball on a trampoline, the more massive the object, the larger the indentation and the greater the area influenced by the object's presence.

If a smaller ball was rolled toward the bowling ball its path would be altered by the indentation created by the larger ball which warps the fabric itself. This, in essence, is the same principle by which the Moon and satellites orbit Earth.

### Straight Lines Through Curved Space-Time

One of the most counter-intuitive notions of general relativity is that objects do not actually curve when they are caught up in a gravitational field. An object like the Moon appears to be traveling in a circular orbit but is actually following a straight path. It is not the path of the moon which bends but space-time itself!

Strange as it seems, this concept should be familiar to us. After all, we live on the curved surface of Earth, where a straight line between two points (between, for instance, New York and London) is not a straight line, but one which follows the curvature of Earth. When referring to a curved space, the shortest distance between any two points is known as a "geodesic" (from the term "geodesy," which refers to the act of measuring the circumference of Earth).

In Einstein's theory, all motion in the universe must be calculated using a new form of geometry which describes the degree of curvature of space-time (which is dependent on the mass of the object which is warping it) and an object's "geodesic" path through this curvature. This becomes monumentally difficult as there are often many different bodies all warping space-time simultaneously. But this is what one must deal with in order to understand the effects of gravity in general relativity.

### SPACE, TIME, AND GRAVITY

Just like in special relativity, in general relativity we must deal with the "flexible" nature of both space and time. Where in special relativity the speed of an object was able to alter its perceptions of distances and the passage of time, the same holds true under gravitation in general relativity. It was predicted by Einstein (and has been shown in experiment) that the influence of gravitation causes time and space dilation effects identical to those caused by motion in special relativity! The two theories are really not all that different—they both deal with motion and with space-time, but in slightly different contexts.

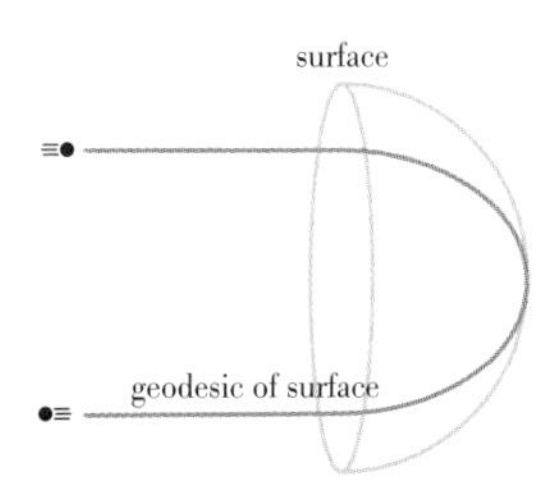

# GENERAL RELATIVITY **Part 2**

The general theory of relativity is widely accepted to be Einstein's greatest accomplishment in the field of physics. With his theory of curved space-time Einstein had rewritten the book on gravity, overturning Newton's greatest discovery, and had given the world the tools with which to use these principles in the form of some fairly complex mathematics. Now, all that was left to do was to demonstrate the validity of his theory and then to use it to learn more about the universe.

## Tests of General Relativity

Though the notion of curved space-time may be difficult to fully understand or believe, Einstein personally came up with several ways by which experimental physicists could attempt to verify his radical new theory.

The first and perhaps most famous of these tests is that of gravity's ability to bend light. According to Einstein's theory of curved space-time, rays of light should have their paths slightly altered when passing near a massive object (and, obviously, the more massive the object the more dramatic this alteration). In 1919, Arthur Eddington successfully observed the light of stars passing by the Sun during an eclipse (the only condition in which such a measurement was possible), which helped to show that Einstein's theory was no hoax. With that discovery, Einstein's fate as the world's most famous man of science was sealed. He was the man who had changed all that we know about the very fabric of the universe.

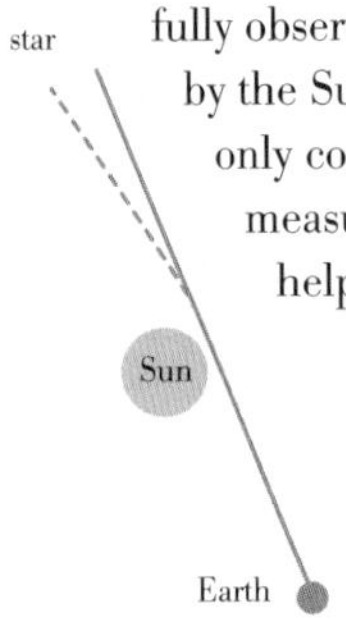

A second test of the theory regarded irregularities in the orbit of the planet Mercury, which had for centuries been an oddity to astronomers, for it could never quite be explained. Newton's equation wasn't able to account for the dramatic curvature of space-time which occurred near the Sun which only Mercury was close enough to truly be affected by. After applying his own equations to the problem, Einstein realized that general relativity presented a much more agreeable solution to this problem.

General relativity, therefore, was not only able to explain all previously understood phenomena, it was able to finally solve previously unexplained occurrences—one of the fundamental tests of a good scientific theory.

star

Sun

Earth

• When passing through the curved space-time around a star, the path of a light ray is bent, proving that gravity has an effect on light.

## Consequences of the Theory

Einstein himself spent much of the decades following his creation of general relativity attempting to use it to understand the behavior of the universe. He was not immediately successful at this, as he believed so strongly that the universe was neither expanding or contracting (even though his theory predicted otherwise) that he inserted a term into his equations which would negate this prediction. This term was known as the cosmological constant, and upon realizing his error Einstein called its inclusion the "greatest blunder of my life."

Despite these initial setbacks and the inherent difficulty in using the theory's complex mathematics, general relativity has afforded scientists an entirely new way of looking at the universe around them, and entire branches of science that would have made no sense before the theory, have become major fields of research.

## Relativity and GPS

One of the most important results of general relativity which truly does play a role in the everyday life of many individuals is the creation of a "Global Positioning System"—a complex network of satellites which continually orbit the Earth and tell us precisely where we are on the planet at any given moment using any number of devices, such as cell phones or vehicle navigation gadgets. It wouldn't have been possible but for our understanding of general relativity (and special relativity), without which we would never be able to precisely sync the time and position of all of these satellites—the effects of warped space-time on time and distance would not take long before they threw off the entire system. We can thank Einstein that today we can know where we are on the planet at any moment.

### WORMHOLES IN RELATIVITY

Some physicists have come to the conclusion that the curved nature of space-time might actually make it possible in the future to do things often relegated to science fiction. If space-time was to actually curve dramatically enough that two distant points actually come close together (imagine bending a piece of paper so that the two far ends almost touch), then it might be possible to find a "wormhole" from one point to another distant point. Doing this would potentially allow us to travel rapidly from one point in the universe to another instantaneously, both in time and space. Don't hold your breath, though—this idea still only exists in the minds of particularly creative physicists.

# Elevators and the Bending of Light

## THE PROBLEM:

Otis, the operator of a "space elevator"—a box traveling through space at extremely high speed transporting people from one planet to another—decides to perform a little science experiment during one of his journeys. He decides to test the effects of his motion on a ray of light, for he knows that Einstein's general theory of relativity states that acceleration through space is equivalent to gravitational acceleration. Therefore, if Otis can show that the motion of his elevator has an effect on light, then gravity should also affect light. How should Otis go about setting up this experiment, and what impact do the results have on how we see light?

## THE METHOD:

Otis decides on a very simple way of setting up this experiment. While the elevator is stationary, he makes a small hole in the side of it which is facing the Sun (his source of light), then he turns off the light. This allows a narrow beam of light to enter the elevator, focused like a laser beam and leaving just a tiny dot of light on the opposite wall, directly across from the hole (Otis measures the distance from the floor of the elevator to the hole and from the floor to the dot of light, finding the distances to be equal).

Now, Otis fires up his space elevator and begins accelerating through space at unimaginable speeds. The acceleration pushes him against the floor of the vehicle—a force identical to the force of gravity.

Otis then uses this opportunity to test the effects of acceleration on this narrow beam of light as the elevator accelerates faster and faster.

## THE SOLUTION:

As the elevator's acceleration increases (a phenomenon identical to an increase in the force of gravity, which is why this is known as g-force), Otis notices that the dot of light appears to move downward. Acceleration affects the path of the light.

The reason for this is simple: in the brief time it takes for the light to travel from the hole to the opposite wall, the elevator has already moved forward slightly (though it would have to be moving rather quickly for this effect to be at all noticeable). Because of the motion of the elevator, the beam of light appears to "bend" as it enters the elevator. Now, carrying this thought through to its conclusion—Otis recalls that there is no way of knowing if the force he feels is caused by the elevator's acceleration or by gravitational force, the bending of the beam of light appears to be caused by gravity.

The implications of this fact are of particular interest: if light, as had been the opinion of physicists for some time, is massless, how could it possibly be affected by gravity? After all, the force of gravity—so said Isaac Newton—is directly dependent upon the mass of two objects. Thus, light should not be bent by gravity.

General relativity has a different approach to the subject, treating gravity as acceleration, rather than a force. Of course, later experiments (though none using space elevators) showed that not only that Einstein was right in this particular prediction, but also that the rest of his general theory of relativity, including his controversial views about gravity being caused by the curvature of space-time, were correct.

Otis' experiment simply verified Einstein's principle of equivalence and the effect of gravity upon a ray of light.

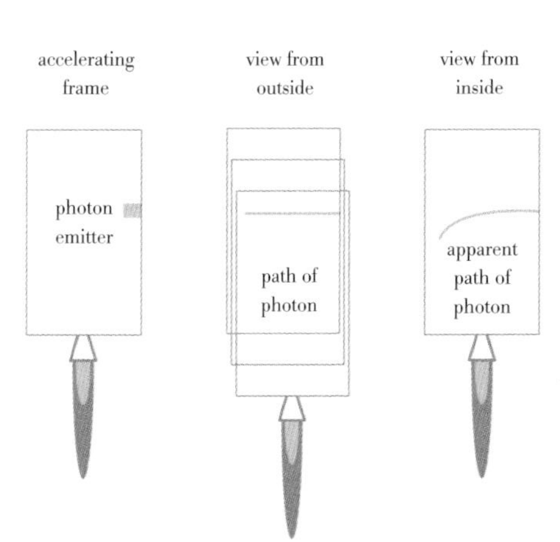


• Einstein imagined an elevator moving through space to explain why light appears to bend in the presence of motion and in a gravitational force.

# QUANTUM MECHANICS

The family of theories in physics which have collectively become known as quantum mechanics certainly seems quite imposing on the surface. These theories deal with elements of the natural world which are far too small to ever be seen, even with the fanciest microscopes (but for a few exceptions). They have been developed to describe the behavior of the smallest particles known to man, and have done so with a very surprising degree of accuracy. The theories in quantum physics have become truly fundamental to almost all of physics today.

**The Importance of the Theory**

Very nearly the entirety of modern physics (both theoretical and practical) is founded on ideas which began in the first decades of the twentieth century under the earliest quantum physicists—hugely important thinkers such as Max Planck, Niels Bohr, Werner Heisenberg, Paul Dirac, Louis de Broglie, and Erwin Schrödinger. The theories of these great scientists ushered in a new era of scientific research.

What has been gained from such studies? What is now known about the universe? How well can we predict the movements and interactions of particles? These are all very complicated questions. In fact, it can be said to some degree of accuracy that the advent of quantum physics has led to a lesser ability to predict motions and interactions. Quantum mechanics does not deal with the behavior of individual particles, but instead focuses solely on probability and statistics. In quantum mechanics we can only ask what a particle (or a group of particles) is *likely* to do, and to calculate as precisely as possible the probability that they will behave in this way.

Quantum physics tells us that within all things—both in the matter of which we are made and the forces which define interactions—there is an inherent "unknowability" that can never be superseded. Yet these theories have led us to the deepest understanding of the physical universe yet achieved.

"For those who are not shocked when they first come across quantum theory cannot possibly have understood it."—***Niels Bohr***

## The Founding Principles of Quantum Mechanics

Quantum mechanics, though undeniably complex, rests on a few key concepts:

### Quantization

When something is broken up into finite, discrete quantities (such as photons of light or individual material particles such as electrons or protons), it has been quantized. In quantum mechanics everything, even the forces themselves, are best understood as the interactions between individual particles.

### Wave-Particle Duality

Isaac Newton theorized that light consisted of particles. In the nineteenth century Thomas Young overturned this theory with his belief that light consisted of waves. With the onset of quantum mechanics and the work of Max Planck, it became clear that light consists, oddly enough, of both. Two decades later, physicists would begin to realize that all particles behave in a "wave-like" manner. According to quantum mechanics, however, these waves are not actual, physical waves, but waves of probability, defining the chances of finding each individual quanta (photon) at any given moment. This concept is known as wave-particle duality.

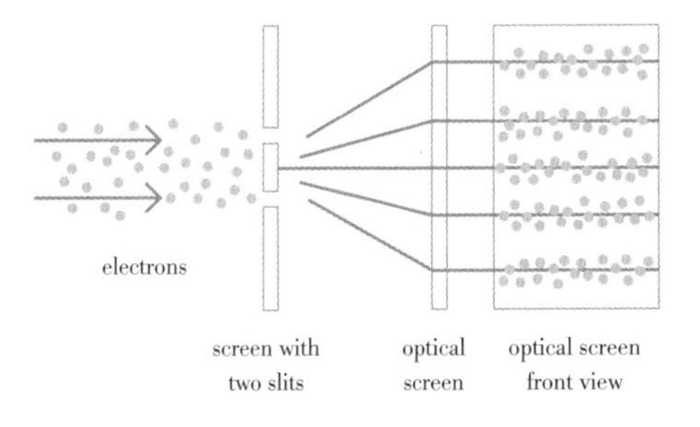


• Wave-particle duality is perhaps best demonstrated in the "double slit" experiment, where individual light photons are passed between two slits and yet interfere with each other as if waves.

### The Heisenberg Uncertainty Principle

This defines the notion that a quantum particle can never be observed as both a particle and a wave at the same time. In other words, the wave qualities (which define the particle's momentum) and the particle qualities (which define the particle's exact position) can never be known at the same time. Thus, one can measure either a particle's position or its momentum, but never both. Thus, a quantum particle is impossible to fully define (hence, the uncertainty).

## Final Observations on Quantum Mechanics

On the surface, these concepts may not appear particularly difficult to understand, yet quantum physics remains a difficult subject, even for physicists. The notion that there are limits to what we can know about the material world is only the beginning of the philosophical and metaphysical difficulties which have led such prominent physicists, including Max Planck and Albert Einstein (two of the founders of these theories), to recoil in disbelief.

For now, it is the best that science can offer; there has not yet been a theory or group of theories capable of toppling the imposing fortress of quantum physics.

Chapter

# Quantum Mechanics

Quantum physics has provided a foundation for some of the most important discoveries in the history of science. Although now more than a century old, it continues to dominate the thoughts of physicists. In this chapter, we look at concepts such as Heisenberg's uncertainty principle, antimatter, and the seemingly paradoxical quantum mechanics, as well as the ways in which these theories have been put to practical use.

**Profile** # Niels Bohr

**Niels Bohr would become a legend in his native Denmark—one of their most important exports and their most intriguing character since Hamlet. The height of his worldwide fame and success would come in the 1920s and '30s with his grand elucidation of the mysterious world of quantum mechanics and his equally grand public persona—but that was well after the world had already begun to accept the truth of a non-classical universe.**

## The Life of Niels Bohr

Bohr was born in Copenhagen, Denmark, in 1885. His father, Christian, was professor of physiology at the University of Copenhagen (he had a medical phenomenon, the Bohr effect, named after him) and his younger brother, Harold, was a mathematician. Clearly, there was a certain amount of intelligence which ran in the family, and Niels was no exception.

In 1903 Bohr enrolled, as had his father (and as would his brother), at Copenhagen University, initially studying philosophy and mathematics. His philosophical background would have a clear impact on his later work, especially as he later became one of the first (and certainly the most eloquent) to begin to describe the more ethereal properties of quantum mechanics and the epistemological implications that these evince.

In 1905, Bohr conducted a series of experiments to examine the properties of surface tension and wrote an essay on the subject, which won a gold medal in a science competition, encouraging Bohr to take on the study of physics instead of philosophy. In 1911 he received his doctorate and moved on to do research in England, first under J. J. Thomson at Trinity College, Cambridge, and then under Ernest Rutherford at the University of Manchester. It was under Rutherford's guidance that Bohr published his quantum model of atomic structure in 1913.

Niels Bohr married Margrethe Nørlund and together they had six sons, four of whom would live long, successful lives (and one of whom, Aage Bohr, would himself win the Nobel Prize in Physics in 1975). In the course of the 1920s Bohr became one of the world's leading spokesmen for the new quantum theories which were being developed by his friends and colleagues throughout Europe. He defended the validity of the theories against the likes of Erwin Schrödinger and Albert Einstein—in fact, Einstein's arguments against the theories of quantum mechanics were directed specifically at Bohr, who took the time to counter them all, thus furthering the scientific dialogue and encouraging others to question and to learn.

"We must be clear that when it comes to atoms, language can be used only as in poetry."—***Niels Bohr***

Bohr won the Nobel Prize in 1922 for his work in developing the new atomic model and during World War II assisted American scientists with the atomic bomb project in Los Alamos. He lived until 1962 when he passed away in Copenhagen at the age of 77.

## Bohr's Science

Niels Bohr's most important scientific work began with his collaboration with Ernest Rutherford at the University of Manchester. Studying under the inventor of the modern atomic model gave Bohr a unique insight into the world of subatomic physics. He had not been around long enough to remember a time before the modern atomic models; he was not alive when the electron was first discovered and he probably could not even have remembered much prior to the earliest quantum physics. Bohr was one of the first of the "quantum generation," and the first to explore the field to anything like its fullest degree. He was determined and single-minded enough that even criticism from the likes of Albert Einstein could not dissuade him from pursuing it even further.

Bohr was the first to apply quantum mechanics to the atomic model, theorizing that as electrons orbit the atomic nucleus they are confined to specific energy levels which are defined by their absorbing and emitting quanta of light (photons). Later, in encouraging the younger generation of physicists to pursue the principles of quantum mechanics, he played an essential role in guiding Heisenberg toward his famous uncertainty principle. Bohr's work in formalizing the principles of quantum mechanics led to the most widely held interpretation of the theory being named the Copenhagen interpretation in reference to his home city.

Perhaps better than any other physicist in the early twentieth century, Niels Bohr embodied the science that would become quantum mechanics, for it seems that he alone had the courage to truly explore this new world and accept what he found there.

### THE BOHR-EINSTEIN DEBATES

Throughout much of the 1920s, '30s, and '40s, Niels Bohr participated in one of the great scientific rivalries in modern history. When Albert Einstein expressed his disagreement with the probability-driven quantum physics of which Bohr was a founding father, several decades of friendly debate ensued. Albert Einstein famously developed a series of thought experiments with which he hoped to question the fundamental tenants of quantum physics. Rather than weakening the theory, however, Einstein's questions drove Bohr and the other quantum physicists to further clarify and strengthen their ideas. It demonstrated to future generations how science is edified through spirited debate.

# THE QUANTUM ATOM **Part 1**

While Ernest Rutherford had done a great deal to finally put the pieces of the atom together, deducing that these tiny little chunks of matter consist of a positively charged nucleus orbited by negatively charged electrons, physicists were still left with several important, unanswered questions. It would take the combined work of Rutherford and Niels Bohr, working with a brand new set of tools (quantum physics) to finally begin to solve these problems and create a new and improved model of the atom.

### The Energy Problem

What exactly were the issues which arose as a result of Rutherford's simple solar system model of the atom? Perhaps the most egregious was the problem of an electron's energy, and the fact that under the laws of physics then understood, atoms should have been positively unsustainable.

It was clear that electrons orbiting the atomic nucleus made use of different energy levels, and that in order to change energy levels they needed to either absorb or emit light (photons). When a photon strikes an electron it absorbs the light and becomes more energetic, jumping to a higher energy level around the atom (these are often called energy shells). Very quickly, however, an electron will spit that energy back out, falling back into a lower energy level.

The problem with this is that the electron, being drawn toward the positively charged nucleus, would be drawn toward the state requiring the least amount of energy, so it should continue to emit energy, falling into smaller and smaller orbits around the nucleus until these two bodies collided. All of this would happen very quickly. In other words, this atomic model was highly unstable.

### Bohr's Solution

The quest to find a stable atomic model led Niels Bohr to his first great contribution to atomic theory: he argued that this question of atomic stability might be solved simply by applying the already widely-accepted Planck/Einstein quantum ideas to the behavior of electrons.

Because Einstein had shown that light exists only in quantized form, this concept clearly had relevance to the atomic structure, for if electrons are known to both absorb and emit light, he knew that they could only do so by way of these individual photons, in very specific, quantized amounts. An atom could not emit a partial photon, for such things do not exist.

The principle can be summed up rather simply: as the energy being emitted by an electron can only come in specific amounts—determined by the frequency of the light emitted, which is unique to each atom—an electron's orbit can only exist at very specific distances from the atomic nucleus. The precise sizes (energies) of these orbits differ with each element because the attractive, or repulsive, forces vary with the size of the nucleus. The larger the nucleus, the greater the energy of the first possible orbital and every subsequent orbital.

The absorption and emission of light by electrons may only change the electron's orbit by exactly the amount of energy contained within a single photon. Bohr used the data which had been gained by experimentation on hydrogen atoms (showing the movement of electrons both before and after stimulation by photons) and found that this change did indeed correspond precisely to the amount of energy denoted by Planck's constant—the number which defines the size of a photon of light. The orbits of electrons were determined by the energy carried by photons, and were thus subject directly to quantum rules.

Bohr explained why it was that electrons never collide with the atomic nuclei: they may only travel at specific energy levels and once they reach their lowest possible energy level (their ground state) they can go no lower except by emitting a "partial" photon, which is a quantum impossibility. A new model of the atom had been successfully created!

## ATOMS AND COLOR

This whole notion of electrons "jumping" from one energy level to another actually enables us to understand a very basic everyday phenomenon: color. Thanks to Niels Bohr we know that certain types of atoms possess arrangements of electrons which absorb or emit light at very specific wavelengths (which we perceive as light of different colors). When light of any color strikes an object, the various atoms in that object absorb specific colors of that light (depending on their electron configuration) and then emit the light back out again. It is the color of the light being emitted by these electrons which we perceive as color. This is why different elements appear different colors to us.

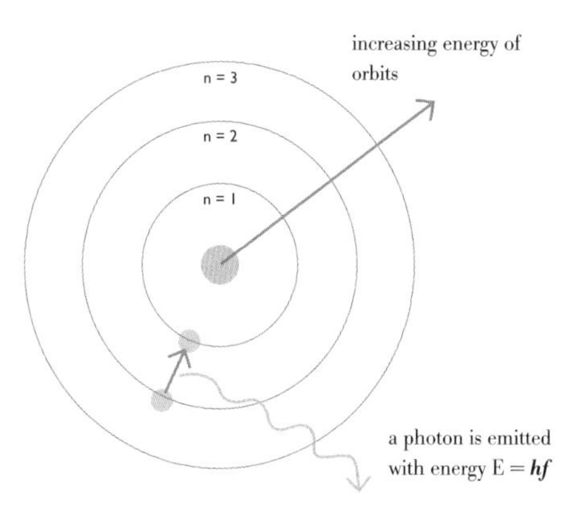


• Bohr's atomic model explained that electrons may only orbit the nucleus in very specific "shells," moving from one to another by absorbing or emitting photons of energy.

# THE QUANTUM ATOM Part 2

Although Niels Bohr had successfully solved the most glaring issue of the atomic structure by applying the principles of quantum physics, other questions would still continue to crop up over the following decade which demanded answers. Fortunately, the answers would arrive almost as quickly as the questions could be asked, and physicists were thus led to an even greater understanding of atoms and the subatomic particles of which they are composed.

## Electron Spin

By the 1920s, a concept had been developed known as quantum numbers that began to explain the behaviors of electrons. These variables—three of which were recognized at the time—defined every aspect of an electron's orbit within an atom as it was then known. Any given electron, it was thought, could be fully explained simply by providing a value for each of these variables. Furthermore, by knowing the precise set of numbers for the valence (outermost) electrons of a given element, all of its chemical characteristics could be determined.

It was soon realized that defining an electron using just three numbers simply didn't explain everything. It was not exactly wrong, just incomplete. The only way to solve this problem was to find a fourth quantum number. This number, first described by the German-American physicist Ralph Kronig in 1924, was given the somewhat playful name "spin." We know now that all electrons (and other particles) possess spin in one of two varieties—spin up and spin down.

The spin of an electron is not the same thing as, say, planetary spin, or the spin of a basketball on one's finger, however. Instead, it is considered to be a measurement of the particle's intrinsic angular momentum—a quality which has less to do with the physical motion of the particle in its orbit and more to do with a property intrinsic to the electron

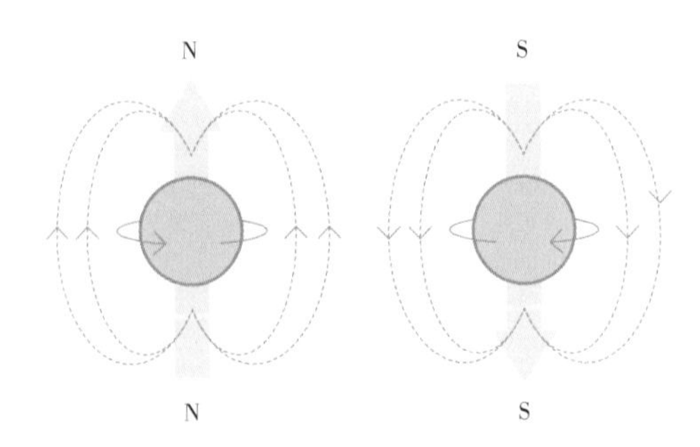


• Electrons are seen to "spin" in one of two directions: up or down.

itself, and which is terribly difficult to explain as there really is no analogy which fully does it justice. We simply know that it exists and without it we would have no hope of fully understanding electrons.

### The Exclusion Principle

Within just a year of the addition of spin to the list of quantum numbers, many, including Austrian physicist Wolfgang Pauli, had become thoroughly convinced of its necessity. With this addition to the list of quantum numbers, physicists could begin to find an answer to a question which had been asked since 1913: What defines the number of electrons in each energy level of an atom? Why can each orbital hold only a certain number of electrons and no more?

In 1914 English physicist Henry Moseley used a spectroscope (an instrument which measures the distribution of color emitted by electrons) to measure electron configurations within atoms and first recognized the fact that electrons arrange themselves only in very particular and predictable configurations, unique to each element.

Bohr's model of the atom led to these crucial insights and provided the first steps toward understanding the behavior of electrons within the confines of an atom, but at the same time it did very little to help understand just why electrons configured themselves as they did. The answer finally arrived with the important work of Wolfgang Pauli in 1925.

Pauli's best-known achievement came in the middle years of the 1920s, with his exclusion principle. Its premise can be stated plainly enough: no two particles may ever be allowed to occupy the same quantum state at the same time.

### What Does This Mean?

When applied to electrons, this principle simply says that no two particles within an atom may ever share identical quantum numbers. If there are two electrons in the same energy level within an atom, they might have different orbital shapes; if they have the same orbital shapes, then they might have different directions of angular velocity. If all of these features are the same, however, we now know that the two electrons must differ in their spin number.

It is now a fundamental, generally accepted truth within physics (and one which is regularly used by chemists in their calculations) that different energy levels within an atom may hold very specific numbers of electrons as a direct result of these atomic numbers. This exclusion principle is a fundamental property of subatomic particles—one which finally gives a full explanation of why the periodic table works as it does and why some chemicals specifically combine with others. Nearly all of modern chemistry has its basis in understanding quantum numbers and electron energy levels within atoms.

# WAVE/PARTICLE DUALITY

In the earliest quantum theories, light could best be seen as individual particles. At the same time, these particles clearly exhibited the behaviors of waves within the confines of certain experiments. Depending on the experiment, either of these very different forms could be seen as being valid, so which is it? As hard as it may be to believe, in the 1920s we began to realize that something can be both a particle and a wave.

**Waves of Matter**

In showing that the waves which formed electromagnetism could also be seen as particles, quantum theory suddenly blurred the line between these two ideas. If there is no distinction between matter and wave, then what classical ideas are left for physicists? Where could a line be drawn between one thing and another?

Difficult as it may be, it does seem to be the case that in quantum theory something can be two things at once.

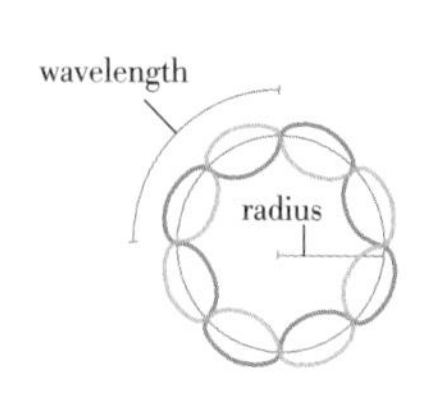


electron orbit = four wavelengths

• Electron orbits within an atom are determined by specific wavelengths of light—only those orbits permitting whole wavelengths of light are allowed.

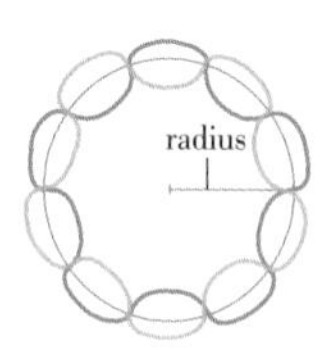


electron orbit = five wavelengths

This train of thought led French physicist Louis de Broglie to theorize that not only light but even matter itself may exhibit the properties of both particles and waves.

De Broglie worked out mathematically all of the factors involved in electron orbits that had been established over the years—their energy, their momentum, the shape of their orbits—and what he found was a new theory of electrons, one which provided a clear physical meaning to the ambiguity embedded into Pauli's exclusion principle.

De Broglie recognized that the most current atomic theory was still rooted in a distinctly classical perspective, where electrons were "forced" into the preconceived image of little balls of matter spinning around other, larger balls. The classical action/reaction dogma still reigned. De Broglie reasoned that because even electrons could be viewed as waves then the atom might actually be far easier to explain in a satisfactory way. The obstacle, of course, was that it meant getting over some stubborn preconceptions about the way things work.

If an electron is viewed as a wave, then the distance at which each electron orbited the nucleus of an atom would correspond to the size of its wave, and only orbits in which whole waves would fit could exist. Electrons couldn't possibly occupy any other orbits because as it gained and lost quanta of energy (in the form of photons), its wavelength would either increase or decrease accordingly. The entire notion of specific electron energy levels was readily explained in this new model.

## A Mathematical Foundation

While De Broglie was certainly following an important and revolutionary path in terms of his ability to conceptualize an entirely new explanation of electron orbits, his theory lacked a new system of mechanics, analogous to the laws of Newton or Maxwell, which would fully govern all particles subject to this quantum behavior and allow physicists to actually make predictions using these new ideas.

The Austrian physicist Erwin Schrödinger began his own work on a wave theory of particles at roughly the same time as De Broglie, but did not publish it until a year later. In January 1926, Schrödinger expanded further upon this enticing theory of wave/particle duality, offering the world his crowning achievement—the Schrödinger equation. It was an equation which asserted first that all matter—not just electrons—could be seen as having an associated wave function. This rather complicated and far-reaching piece of mathematics held the incredible power to provide a consistent method by which the given state of any quantum system at any given time could be fully calculated.

Schrödinger's equation can be credited as one of the most important moments in twentieth century physics, serving to provide the first solid mathematical footing upon which the debate over quantum physics in the 1920s and '30s would rest. The equation allowed physicists studying the atomic world to finally predict with some accuracy the quantum behavior of subatomic particles, and to explain the phenomena they had been observing.

### WAVES OF WHAT?

Perhaps the most important aspect of the Schrödinger equation was that it sought to remove entirely the concept of quantum mechanical objects (such as photons, electrons or any other particle) being real, physical objects at all. The waves in quantum mechanics are not physical waves at all, but waves of pure probability, which allow us to measure the chances, at any given time, of finding a particle at a certain place. Though inadvertently, Schrödinger's work led to some of the most difficult questions about what matter actually consists of—perhaps even we ourselves are nothing more than probability!

**Profile** # Werner Heisenberg

**Werner Heisenberg infused the world of physics with uncertainty. Though perhaps unintentionally, his work helped to define the very limits of what human beings can and cannot know. A brilliant quantum physicist, Heisenberg helped to found the Copenhagen interpretation of quantum mechanics, which continues to aid physicists even today in defining the methods and meanings of quantum physics.**

## The Early Life of Heisenberg

Werner Heisenberg was born on December 5, 1901, in Würzburg, Germany, the son of Dr. August Heisenberg (a professor of Greek at the University of Munich) and Annie Wecklein. As is fitting for the man he would become, Heisenberg was a child prodigy, mastering the piano at a young age and driven to the pursuit of science by his father's encouragement. As a teenager he taught himself calculus (not a particularly easy task) and became heavily engrossed by the challenges presented by physics.

Heisenberg attended the Maximilian school in Munich before attending the University of Munich beginning in 1920, where he studied theoretical physics under the famous Arnold Sommerfeld and alongside Wolfgang Pauli. During his time at the University he traveled to Göttingen to study under, among others, Max Born, who would provide considerable aid to Heisenberg in developing his theories.

Heisenberg received his PhD from Munich in 1923 and returned to Göttingen to assist Max Born. Also in the middle of the 1920s Heisenberg worked with Niels Bohr (see pp. 128–129) at the University of Copenhagen.

## Matrix Mechanics and Uncertainty

The decade of the 1920s was the most fruitful of Heisenberg's scientific career, thanks in part to a bout of hayfever in 1925 which saw him move to an island called Heligoland just north of the German coast in order to recuperate. There he was given the time and space needed for him to develop the initial thoughts which would lead him to the development of an entirely new and complete system of quantum mechanics known as matrix mechanics. This system used the relatively unknown mathematics of matrices in order to solve quantum mechanical problems.

Heisenberg published his paper, alongside Max Born and Pascual Jordan, in 1926. The following year these new methods of calculation

> "What we observe is not nature itself, but nature exposed to our method of questioning."
>
> ***—Werner Heisenberg***

led Heisenberg to perhaps his most memorable achievement—the Heisenberg uncertainty principle (see the following page for details), which defined the limits of human knowledge in regard to quantum particles. That same year Heisenberg was appointed head of the physics department at the University of Leipzig, where he would remain well into World War II. By the early 1930s, even though still a relatively young man, Heisenberg was world famous and hugely respected in the physics community. He traveled the world during those years and in 1932 he was awarded the Nobel Prize in Physics for, "The creation of quantum mechanics…" Certainly a statement of great honor for a physicist in those times.

## The Heisenberg Controversy

In 1941, with World War II in full swing, Heisenberg was appointed Professor of Physics at the University of Berlin and Director of the Kaiser Wilhelm Institute for Physics there.

While many physicists (especially those of Jewish heritage, such as Einstein and Born) had either fled Germany or been forcibly removed from their positions by Nazis, Heisenberg remained loyal to his nation and to the Nazi party, forever marring his standing among the world's scientists.

There is much debate to this day over precisely what kind of work Heisenberg performed for the Nazis during the war, especially in regard to their atomic bomb program. While Heisenberg claimed to have actively stalled the bomb program in order to prevent Hitler from creating a bomb, there is also evidence that this may not have been the case.

• At the Fifth Solvay International Conference, held in Brussels in 1927, the world's leading scientists gathered. Werner Heisenberg can be seen on the back row, third from the right.

## After the War

After the Allied victory, Heisenberg was arrested and brought to England along with other German scientists. Though his standing had been tarnished due to his Nazi involvement, not all his friendships were ended. Most notably, perhaps, was Paul Dirac (see pp. 140–141), who gave his old friend Heisenberg the benefit of the doubt with regard to his wartime activities.

Heisenberg returned to Germany in 1946 and was appointed director of the Max Planck Institute for Physics and Astrophysics at Göttingen, where he remained until 1970, even after the institute was moved from Göttingen to Munich.

Toward the end of his life Heisenberg continued his work, particularly in the fields of thermonuclear physics and in the all-important quest to find a unified theory of elementary particles—a quest which continues to this day and which is called the standard model of particle physics. Werner Heisenberg died of cancer on February 1, 1976, at the age of 74, having almost entirely repaired his reputation from the war.

# THE HEISENBERG UNCERTAINTY PRINCIPLE

In philosophy, epistemology is the study of knowledge. It is the broad, open ended search for answers to such daunting questions as, "What is the nature of knowledge?" and, "Just how much are humans capable of knowing?" Humans have long believed that it was in their power to know everything. The Heisenberg uncertainty principle forces us to ask a question which almost seems more philosophical than scientific: What can we know? And the answer is disturbingly simple: Not nearly as much as we thought.

**The Formulation of the Principle**

Werner Heisenberg was the first to inform the world that determinacy—a belief that the universe is predictable and operating under wholly consistent laws—simply does not exist. Within quantum mechanics there are things that we simply cannot know, for nothing is determined. The key to Heisenberg's ultimate achievement was the mathematical method he helped to create. Using this tool, a consistent method for solving quantum problems could be found, where prior to this the only way to solve problems at the quantum level was to begin with the equations of classical mechanics and then to "tweak" them in such a way that the desired answer might be achieved.

**Complementarity**

The uncertainty principle is founded on a concept developed by Niels Bohr known as complementarity. This concept asserts that certain measurable properties of particles come in pairs which are inextricably linked. These properties are so thoroughly intertwined that the measurement of one property has a direct result on the measurement of the other.

The most frequently cited example of this principle in action is the problem of measuring the position and momentum of any given particle at the same time—two complementary properties which, according to the uncertainty principle, can never be known simultaneously. We can know one property or the other, but we can never achieve exact measurements for both. We can discover where an electron is (its position), but we will lose all knowledge of its speed (momentum), and vice versa.

This principle is essentially nothing more than a logical result of wave/particle duality. An electron can exist as

## THE PHILOSOPHY OF UNCERTAINTY

The uncertainty principle has over the decades led to an abundance of philosophical questions regarding what exactly quantum mechanics actually means. Questions raised, even by physicists such as Niels Bohr, are certainly intriguing: If something can be both a particle or a wave, then is it really anything at all? If the existence of a particle depends upon our measurement, then does the particle really exist in the first place, or do we cause it to exist by measuring it? What kind of consciousness is required to bring matter into existence? And of course there are many more questions beyond these, all directly resulting from the uncertainty principle.

either a wave or a particle, but never both—some experiments are meant to study an electron as if it was a particle while others study it as if it was a wave, but no experiment may treat an electron as both a particle and a wave at the same time. When we attempt to find the position of an electron, we are thinking in terms of particles (for a wave isn't readily associated a single position), while momentum is a measurement of an electron in motion (as a wave). It cannot exist in both forms at the same time.

### The Results of the Principle

While the uncertainty principle was not immediately accepted within the scientific community and was even treated with great skepticism by some (Einstein, for one, never fully accepted the theory), Heisenberg had offered a quite revolutionary version of quantum mechanics which made a complete break from the past. It enabled him to solve problems that were made more complex by this greater understanding of the relationship between waves and matter.

Heisenberg's theory of quantum mechanics was complicated by the fact that it did not operate according to the older mechanical explanations of either particles or waves. Most importantly, it refused to allow precise calculations for such things as the positions of electrons as they orbited atoms. Instead it provided approximations whereby the electrons' position was given as an area corresponding to the shape of these waves.

Heisenberg's work led both to increased understanding of the behavior of quantum particles as well as the complete, utter absence of certainty.

**Profile**

# Paul Dirac

**There are few physicists who have shaped the way we look at the sub-atomic world as dramatically as Paul Dirac. Though many of the stories involving Dirac focus on the peculiarities of his personality (which are certainly many), Dirac's direct contributions to the study of physics led at long last to a final understanding of electrons (and all other particles) as quantum entities and to the prediction (and eventual discovery) of antimatter.**

## The Education of Paul Dirac

Paul Dirac was born in Bristol, England, on August 8, 1902. His father, Charles, was Swiss by citizenship and taught the French language in Bristol. His mother, Florence, was English. Paul was the middle of three children, and though there is some dispute about his childhood, it appears that life in the Dirac household was not easy, as Paul blamed much of his peculiar personality later in life on the actions of his father.

Dirac attended Bishop Primary School in Bristol and even then showed a remarkable ability in the mathematical subjects. At the age of 12, Paul attended the secondary school where his father taught. Like his brother Felix, Paul went on to study electrical engineering at the University of Bristol. His background in electrical engineering would later provide Dirac with a unique perspective on the world and affect the methods by which he performed his physical research. Dirac obtained his engineering degree in 1921, but failed to find a permanent position in this field. Seeking to pursue his love of mathematics, Dirac began attending Cambridge University in 1923.

While at Cambridge, Dirac became interested in the new theories of relativity and began to serve as a research associate, immediately proving himself to be exceptionally capable at both research and theory. Dirac's first notable foray into quantum mechanics came with an early analysis of Heisenberg's uncertainty principles in 1925, which Dirac recognized as an example of noncommutative algebra (a lesser-known field of mathematics he had studied extensively) and essentially rewrote

• A young P. A. M. Dirac at work at his blackboard, attempting to create "beautiful" equations.

## THE PUREST SOUL IN PHYSICS

Dirac's personality is nearly as legendary as his science. There are scores of tales about Dirac which exemplify his quiet, introverted demeanor—it is said that he always spoke in the most literal and concise language. In one instance, after one of Dirac's lectures a student chimed in that he did not understand one of the problems. After being chided for not answering the question Dirac finally responded: "That was a statement, not a question."

Heisenberg's principle using the language of mathematics. This work led to his PhD in 1926, after which he traveled to Copenhagen to work with Niels Bohr and then on to Göttingen to work with Robert Oppenheimer and Max Born, among others.

## The Dirac Equation and the Anti-Universe

In 1928 Dirac performed his most important work, combining the principle of relativity with quantum mechanics and creating of the Dirac equation. The Dirac equation became a tool of incalculable worth for physicists to predict quantum phenomena. Using this equation, Dirac predicted the existence of a new form of matter—a particle which was identical to the electron, but with an opposite charge. This would become known as the positron. Though contemporaries tended to scoff at the bizarre assertion that an "antiparticle" existed, Dirac went on to propose that not just the electron but all particles possessed "anti" counterparts.

Within just four years the first positron was discovered and Paul Dirac's work was completely vindicated. Antimatter did indeed exist and Dirac's work won him the 1933 Nobel Prize. Carl Anderson, discoverer of the positron, won the prize for his own discovery in 1936.

## The Principles of Quantum Mechanics and Beyond

In 1930 Dirac published a textbook called *The Principles of Quantum Mechanics* which described his own views on quantum mechanics. This work was a monumental achievement, supplanting all current textbooks on the subject, and it remains still very much in use today. That same year saw Dirac elected a Fellow of the Royal Society. Two years later, Dirac was appointed Lucasian Professor of Mathematics at Cambridge, followed by the Nobel Prize the following year.

While at Cambridge, Dirac changed direction slightly and published papers on cosmological topics and for a time during World War II he worked in atomic physics, using the engineering skills he had developed at Bristol University to develop methods of separating uranium for use in both atomic weapons and atomic energy.

In 1969 Dirac retired from Cambridge and moved his family to the United States to serve at both the University of Miami and at Florida State University, where in 1971 he was appointed professor of physics. Dirac died in Tallahassee, Florida, in 1984. He is commemorated by a plaque in Westminster Abbey denoting the Dirac equation.

# ANTIMATTER

In 1928, Paul Dirac developed his most important contribution to physics—the Dirac equation. Not only did this equation play a crucial role in enabling quantum physicists to calculate the behaviors of particles such as electrons, but it led to the prediction of antimatter. Dirac informed the world that for every particle there also exists an "antiparticle." This idea changed the way physicists viewed all matter.

**A Sea of Negative Energy**

Dirac began his journey toward antimatter with a version of Einstein's famous equation of mass-energy equivalence ($E = mc^2$). In the case where momentum must be included in the equation (the equation to determine the relativistic energy/momentum relationship), he began with a relatively simple formula:

$$E^2 = m^2c^4 + p^2c^2$$

Dirac knew that this equation could be solved using two equally viable answers, with the value of E being either a positive or a negative number. In mathematics, both answers should be given equal weight.

This was understood prior to Dirac, but these negative answers were simply disregarded and assumed to have no bearing on reality. The positive value was accepted as truth (for what could possibly be meant by negative amounts of energy?). Dirac took a chance and decided to trust the mathematics over the experimental data, toying with the equation using negative answers and the possible meanings therein.

What Dirac found was rather strange. According to the equation, within an atom there must be both positive energy levels (occupied by normal electrons) and negative energy levels. This led him to declare (rather controversially) that all of that "empty" space within and in between atoms was not really empty at all. It was merely a sea of negative energy electrons.

Dirac went even further, proposing that if this negative energy was supplied enough energy (the required energy can be calculated by Einstein's original equation: $E = mc^2$), a positive energy particle (such as an electron) could be produced from the negative energy. This process would not only create an electron, but would leave an electron-sized hole in the field of negative energy. Dirac proposed that this hole would to any outside observer appear as a normal electron, but with opposite electromagnetic properties (positively charged).

## The Positron and Antiparticles

This hypothetical particle became known as the positron. Dirac had no physical justification for this theory at first—only the mathematics in which he trusted. Dirac has been quoted as saying that, "it is more important to have beauty in one's equations than to have them fit experiment."

Dirac and his equations were finally justified four years after his original proposal, when the first traces of real-life positrons were found by scientist Carl Anderson in 1932.

The positron, however, is certainly not the only antiparticle. Dirac's theory holds true for other pieces of antimatter, no matter how large—antiprotons and antineutrons, for example. Every particle possesses an antiparticle counterpart, and many of these have been found in experiments.

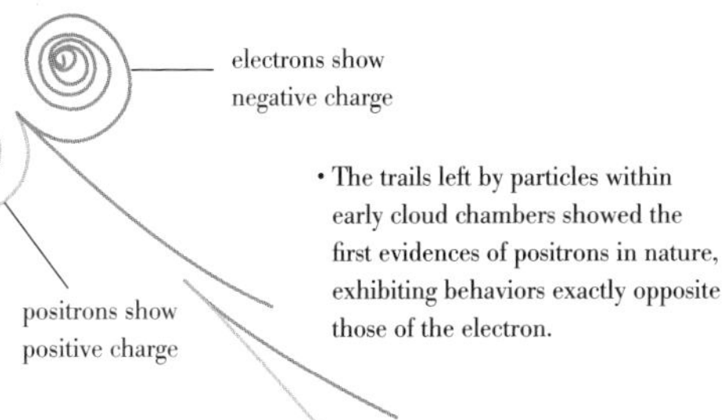


• The trails left by particles within early cloud chambers showed the first evidences of positrons in nature, exhibiting behaviors exactly opposite those of the electron.

## Antimatter in Nature

Though when Dirac proposed the existence of antimatter it was considered to be somewhat exotic, it has become clear to us that these antiparticles do not exist merely within the confines of our experiments. In fact, Carl Anderson's initial discovery of the positron was not in a particle accelerator, but in a shower of cosmic rays—a wholly natural phenomenon. We have reason to believe that a number of different antiparticles are produced by cosmic rays (though these particles are extremely short-lived). We also know that certain antiparticles known as antineutrinos are consistently produced by the nuclear reactions within the Sun (as are positrons), and are rather common in our universe. Perhaps antimatter is not quite as exotic as we thought.

### ANNIHILATION

The existence of antimatter poses an interesting and important question: If it is being produced in nature, why do we not see more of it? There is a simple answer for this and also a complicated answer. The simple answer is that because matter and antimatter are oppositely charged, when antimatter is produced it is very quickly drawn toward matter. The two particles collide and are immediately annihilated, turning into pure energy. This is where the potential for producing energy using antimatter comes in. The complicated answer, on the other hand, goes all the way back to the Big Bang, asking why there is more matter than antimatter in the first place—and in truth we have not yet answered this question entirely.

# PROBABILITY AND QUANTUM MECHANICS

Quantum physics makes it quite clear that a particle cannot exist as both a wave and a particle at the same time and that a particle may possess both wave-like and particle-like properties. But what does this all mean? What is really happening to an electron when it "spreads out" into a wave as it orbits within an atom? Where does the particle go? And what implications does this have on the material world which we call home?

**Waves of Chance**

The answer to these questions begins with an explanation of what we mean when we refer to quantum waves in the first place. With the development of Schrödinger's wave equation (see p. 135) and the principle of uncertainty (see pp. 138–139), it was clear that we must dispense with common sense and accept the reality of the situation: the wave described by Schrödinger is not a physical wave that describes the motion of an electron but rather it is a wave function that describes the probability of finding an electron at a given location at a given time.

In quantum physics we find ourselves describing probability waves rather than actual waves. When we say that a particle moves as a wave, we mean—according to the popular interpretation—a wave of oscillating probability and that if we were to pick a point within that wave at a given moment, we could use Schrödinger's equation to calculate the probability of finding a particle at that location (even if the particle doesn't necessarily exist until we actually look for it). The act of finding a particle in the midst of a probability wave is often referred to as "collapsing the wave function," for finding the particle forces all probability to disappear.

**Making Use of Probability**

So how can we ever hope to transcend this seemingly frustrating aspect of quantum mechanics and actually find some way to use it? The answer is the same as some of the early chemists found when looking at the motion of particles within gases—we must stop looking at the behavior of individual particles and begin looking at the problem as a whole, in terms of statistics and probability.

A familiar analogy to this is the simple flipping of a coin. On any given flip you have a very clear 50% chance of being either right or wrong—an equal chance of either heads or tails. Your chances of guessing each flip individually is not very good. In fact, your chances of guessing five flips in a row are not much better (it is not at all inconceivable that this may result in five heads in a row, or even five tails), but what if you flip the coin 100 times, guessing that about 50% will result in heads and 50% in tails? Statistically speaking, your results should be much more accurate. Now, what if you decided to flip a coin 1,000 times, or better yet, a million? It is a hard and fast rule of statistics: the larger the sample, the higher the accuracy. This same rule also holds true for particles.

Asserting the existence of uncertainty on a microscopic level, therefore, does not necessarily inhibit our ability to make measurements or understand how particles behave. It merely says that when doing so, we are forced to do it in a statistical sense, asking not about the behavior of individual particles (for they could potentially show up anywhere in their probability wave), but about a system as a whole.

By accepting the truth of this principle, some of the greatest advancements in physical theory have been made, from the principles of quantum electrodynamics in the 1940s and '50s to the ongoing efforts to find the ultimate theories which govern everything. This is possible because we have accepted the truth of wave/particle duality and the uncertainty principle. We move forward in physics only by accepting our own limitations and learning to use them to our advantage.

## QUANTUM TUNNELING

One of the most unusual side effects of viewing quantum waves as merely waves of probability is that it allows particles to engage in some rather peculiar behavior. For example, if a probability wave extends beyond an "impassable" barrier, there is a chance that the particle may suddenly "jump" to the other side. This phenomenon is called "quantum tunneling," and lest it be written off as mere conjecture, this fact of nature has been extremely well-documented and even put to use in modern circuitry and in such tools as the "scanning-tunneling microscope."

CLASSICAL PHYSICS

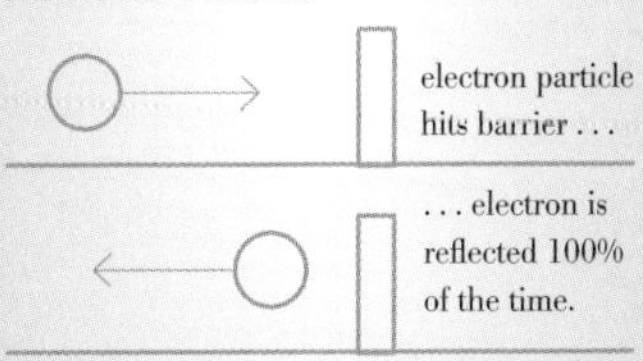


QUANTUM PHYSICS

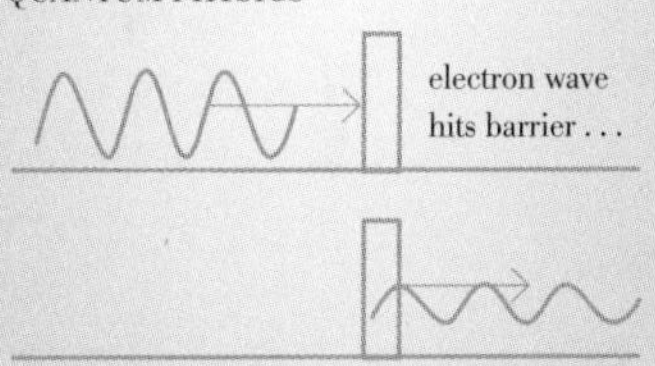


. . . particle is most likely reflected, but there's a small probability that it will "tunnel" through and spontaneously appear on the other side.

# Thought Experiment: Schrödinger's Cat

**THE PROBLEM:**

Erwin Schrödinger is not at all satisfied with the state of quantum mechanics. Though he helped to invent this branch of science, he is aware how difficult it is to believe that any particle could be "spread out" in a probability function, nor that any object could exist as both a particle and as a wave at the same time, existing in some sort of in-between state (called superposition) until it is measured.

Because of this dissatisfaction, Schrödinger sets out to demonstrate to the world just how foolish this idea is. How can he possibly convey the reality of wave/particle duality and superposition in an easily accessible way? How would this affect our understanding of the universe?

**THE METHOD:**

Schrödinger knows that he needs a clever and simple way of demonstrating the strange results of believing that quantum physics consists of ethereal probabilities and dualistic views of matter which lead one to question all of physical reality.

Schrödinger knows that one such phenomenon which physicists believed to behave in this fully probabilistic way is radioactive decay. Therefore, if he can find some way to use the probability of radioactive decay to affect some real world object, he can finally demonstrate the absurdity of the entire theory. With that in mind, he finally comes up with the perfect thought experiment (not to be tried at home):

Put a cat into a sealed box, along with some radioactive element. It is important that the element has a decay rate such that there is exactly a 50% chance that it will emit one particle over the course of an hour (this is a hypothetical element).

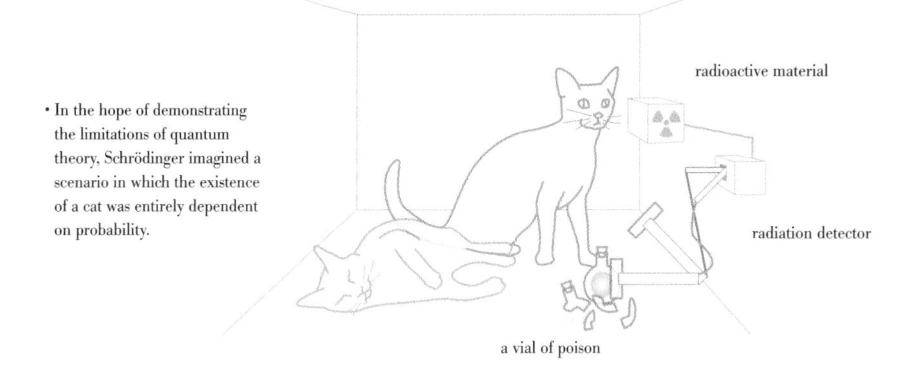


• In the hope of demonstrating the limitations of quantum theory, Schrödinger imagined a scenario in which the existence of a cat was entirely dependent on probability.

Alongside the cat and the radioactive element place a Geiger counter (a device used to detect radioactive decay) and a vial of poisonous gas. Next, rig up a contraption which will break open the vial of gas if the Geiger counter detects a radioactive particle and seal up the box.

## THE SOLUTION:

Because radioactive decay is subject to probability, we cannot predict exactly when a single atom will choose to decay in a given timeframe. So if this experiment is set up correctly, there will have been equal probability after one hour's time of opening the box and finding either a live cat or a dead cat.

The point, however, is not what is observed when the box is opened. There is really not much to learn from either a live or dead cat. Rather, the question is what happens in the box when no one is looking. We want to explore the nature of superposition. Without opening the box and looking inside it is pointless to ask if at any given moment radioactive decay has occurred or not, because this act of decay—a purely quantum event—should be in a state of superposition until the moment the box is opened. The substance both has and hasn't decayed until the box is opened and it is forced to choose one or the other.

Schrödinger is trying to make the point that while the box is closed it is pointless to ask if the cat is either dead or alive because the question itself has lost all meaning. When the box is closed the entire system enters a state of superposition, including the cat, which enters into a state of being neither dead nor alive! As Schrödinger himself puts it: "The psi-function of the entire system would express this by having in it the living and dead cat (pardon the expression) mixed or smeared out in equal parts."

Silly as it may seem on the surface, this experiment became the most famous argument ever devised to attempt to demonstrate the "absurdity" of quantum mechanics. It did not destroy the theory, of course, but it certainly led to many good questions and made people think long and hard about just what these theories meant. If nothing else, this thought experiment demonstrates that in quantum mechanics, reason and logic are simply not to be trusted.

# PRACTICAL QUANTUM MECHANICS

Sure, quantum mechanics is interesting. How could it not be with all its outlandish claims that seem to describe a world other than our own, yet which appears to be true. But is this all there is to it? Is it merely fascinating or is there a practical element to it as well? Is there any point to studying quantum mechanics other than to satisfy our curiosity about the physical world? Just like any other physical science, there is certainly the potential for quantum mechanics to directly affect the world as we know it.

### A Tool for Understanding

First and foremost, quantum physics is beneficial in that it helps us understand how things work. Since everything is made up of atoms, which are held together and acted upon by forces, understanding the principles which govern these things certainly allows us a better understanding of the world around us. But more than this, understanding how quantum mechanics comes into play in everyday life can actually help us to make things better!

Modern electronics in particular operate using quantum mechanical principles. Every object incorporating modern circuitry (which today seems to encompass nearly every object you can find) is essentially quantum mechanical, making use of semiconductors, transistors, diodes and other miniature devices which take advantage of unique principles such as quantum chemistry and quantum tunneling. The truth is that quantum mechanics is everywhere, and we are only just scratching the surface in learning how these principles can lead to better products and better manufacturing.

### A Tool to Study Quantum Physics

Quantum physics helps us to understand quantum physics. Many of the tools we now use to study it are, essentially, only made possible by quantum physics.

For example, one of the only ways we have of actually observing particles is with a device called a scanning-tunneling microscope, which uses the wavelength of electrons in order to visualize an even smaller world than would be possible with even the most powerful optical microscope. In this technology, a metal tip is positioned with incredible precision only a tiny distance (a matter of only a few angstroms—or a few ten billionths of a meter) from the substance (often a metal) being studied. When a charge is then applied to the tip, electrons are drawn from the source and are able to make use of tunneling across this vacuum to the tip. The patterns of this tunneling can then be analyzed by a computer, which translates the data into a visual representation of the atomic structure of the subject.

Beyond merely allowing us some of our first images of actual atoms, these microscopes have enabled us to manipulate individual atoms in order to place them in certain designs and patterns, forming the beginnings of an entirely new science known as nanotechnology—a branch of science which is only just now beginning to bear fruit and promises to deliver new materials of unbelievable strength, not to mention significant advances in medicine.

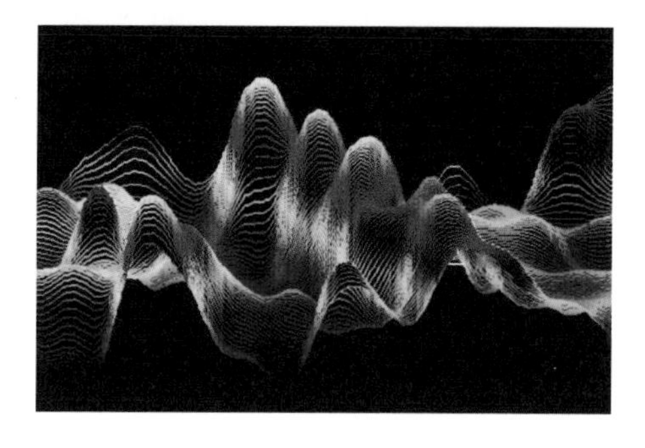

• Images from Scanning Tunneling Microscopes show that atoms do not exist at precise locations, but are spread out as functions of probability.

## Quantum Computers

Advances in quantum mechanics have led to continued advancement in computing power and the advent of greater forms of technology, but they also promise to one day lead to quantum computers.

Where today's computers operate using bits (binary units) of information, quantum computers use qubits (quantum binary units), which hold the potential to work faster and contain far more information by utilizing quantum superposition. Where a bit can only exist as either a 1 or 0, a qubit can exist in a superposition of these two states.

It has been predicted that even a relatively simple quantum computer has the potential to perform operations millions of times faster than today's most modern computers. All that stands in our way now is one of the most difficult engineering challenges ever faced by mankind—to manipulate individual quantum particles in such a precise way that they retain their quantum mechanical features while at the same time allowing the storage and processing of vast amounts of information. Easier said than done.

## Other Uses

While these are just a few of the practical applications of quantum mechanics that are in use or are hovering just beyond the horizon, rest assured that there are many more besides. Quantum mechanics continues to enable incredible advances in research thanks to some new and exciting particle accelerators, amazing advances in the technology of codemaking and -breaking (there is an entire field called "quantum cryptography") and much more!

Chapter

# Modern Physics

This final chapter surveys the state of physics today: What are the most exciting fields of study? How is research in physics currently being performed? And perhaps most importantly, where is physics going to take us in the future? Within this chapter are explanations of the current state of particle physics, an introduction to the expensive but all-important particle accelerators, a look into the mysteries of black holes, and a glimpse into the most promising theories of everything.

**Profile** # Richard Feynman

**Richard Feynman was one of the most brilliant and unique characters in twentieth century physics. Though a Nobel Prize-winner and one of the finest minds of his generation, Feynman was also known for his wit, his curiosity, and his sense of adventure. Whether moving to Brazil and mastering the bongo drum, traveling to Mexico and learning to translate Mayan mathematics, or learning to break into the safes at the Los Alamos nuclear facility during World War II, Feynman's adventures are legendary.**

## Rise to Prominence

Richard Feynman was born in Queens, New York, on May 11, 1918, to Melville and Lucile Feynman. At the age of ten the family moved to Far Rockaway, New York. By his own account, Feynman was greatly inspired by his father and developed a curiosity about the natural world—a curiosity that took deep root. After graduating from high school Feynman was accepted by the Massachusetts Institute of Technology and began studying there in 1935, obtaining his Bachelors of Science degree in mathematics in 1939. Only then did he move into physics.

Leaving MIT, Feynman pursued his graduate work at Princeton University, where he first became interested in Paul Dirac's work on quantum mechanics and began to apply some of his own original thoughts to these theories. Feynman's doctoral work at Princeton led Feynman, at just 23 years old, to begin to think about the question of electron and electromagnetic interactions in a new and creative way. This work culminated in an original and exciting explanation of quantum mechanics known as quantum electrodynamics (see pp. 154–155).

Another of Feynman's most famous inventions was the Feynman diagram. This was a simple way for physicists to graphically represent complex mathematical equations and simplify complicated phenomena. Feynman won the Nobel Prize in 1965 for his work in quantum electrodynamics, sharing the award with Julian Schwinger and Japanese physicist Sin-Itiro Tomonaga.

## World War II and the Manhattan Project

Feynman had little time to savor his PhD before being invited to join the race to build the atomic bomb. Feynman moved to Los Alamos to join some of the greatest minds in physics in an attempt to solve one of the greatest problems in physics in record time.

Unfortunately for Feynman, it was while in Los Alamos that personal tragedy struck—his first

## THE ACCESSIBLE SCIENTIST

Richard Feynman is perhaps the most easily relatable of the prominent scientists of the twentieth century. He published numerous books in his lifetime, several of which are easily accessed by the common reader. Two of these books are filled with humorous stories and scientific anecdotes—*Surely You're Joking, Mr. Feynman* and the sequel, *What Do You Care What Other People Think?* Anyone interested in gaining insight into a peculiar scientific mind, will find these books absolutely invaluable.

wife, Arlene, who had been diagnosed with tuberculosis, passed away at a nearby hospital.

Though grief stricken, Feynman returned to his work, witnessing the first ever test of a nuclear explosion in the New Mexican desert and taking some time here and there to test the security at the research facility by learning to pick the locks of the safes which held the nation's most important nuclear secrets.

• The Challenger Space Shuttle disaster, and the subsequent investigation, placed Richard Feynman firmly in the public eye.

### Later Life and Work

After the war, Feynman was appointed as a professor of theoretical physics at Cornell University and then, in 1950, became a professor of theoretical physics at the California Institute of Technology, which would be home for the remainder of his life. Before taking up his post at Caltech, Feynman spent ten months on sabbatical in Brazil, where he avidly took up playing the bongo drum.

Along with continued work in Quantum Electrodynamics at Caltech, Feynman branched into other subjects, such as a theory of partons which led fellow Caltech professor Murray Gell-Mann to propose that some particles were made up of even smaller quarks (see p.159). By early 1979 Feynman's health had begun to wane and he had surgery for stomach cancer. The 1980s saw Feynman's fame reach even greater heights when he joined the committee set up to investigate the Challenger Space Shuttle disaster in 1986, leading Feynman to demonstrate live on television the failure of the Shuttle's O-rings. Richard Feynman died February 15, 1988, at the age of 69 from a recurrence of stomach cancer.

# QUANTUM ELECTRODYNAMICS

Richard Feynman described his greatest achievement, quantum electrodynamics (QED), in his book *The Strange Theory of Light and Matter*. Strange it most certainly is, but its strangeness is surely overshadowed by its accuracy. It is difficult to overemphasize just how all-encompassing QED is—practically any observable event in our universe must in some way be governed by the laws regarding light and electrons. Every bit of matter of which you and I are made—every force, action . . . everything—comes down to just this: the behavior of light and electrons.

**What is QED?**

So, what unfathomably complicated concepts and sets of laws must one come to terms with in order to even begin to understand such a far reaching theory as QED? What new depths of academic learning must we detangle in order to even begin to wrap our limited brains around this greatest of all academic achievements? How can we define things like light and electrons—quantum elements bound by uncertainty?

In fact, QED involves just three simple actions:

*Action #1*: An electron goes from place to place

*Action #2*: A photon goes from place to place

*Action #3*: An electron absorbs or emits a photon

That's truly all there is to it—just those three steps, none of which introduce any particularly new ideas which have not been discussed elsewhere in this book. There is nothing too shocking. Every action and reaction we observe in the universe can, in one way or another, be simplified to just three primary actions, which form the backbone of QED.

It is only when we actually attempt to put QED to use that we find its true value, both in explaining the material world and in revealing some of its deepest mysteries.

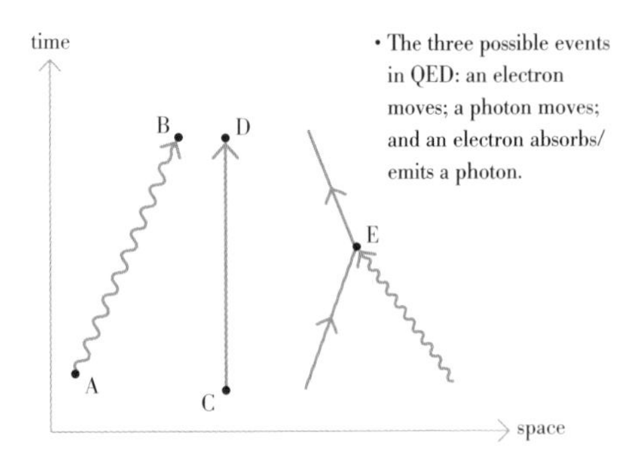


• The three possible events in QED: an electron moves; a photon moves; and an electron absorbs/emits a photon.

## The Sum Over Histories

As stated before (but it deserves to be said again) quantum electrodynamics helps us to explain everything. Want to know just why light reflects off a mirror or why items placed in water appear distorted? Want to know why we see mirages on the road on a hot day? Want to know how lenses work to focus light? How atoms bind together electromagnetically? The answers lie in QED and the three simple actions listed earlier.

Both the value and mystery of QED come from looking at some very peculiar questions regarding these actions. For instance, as an electron attempts to move from one position to another, we must ask if there are any specific laws which might prohibit it from taking anything but the most direct path? In fact there are no such laws. An electron (or a photon) may take absolutely any path it chooses from one place to another, no matter how erratic (or indeed, improbable) it might seem.

In fact, the only way to solve problems involving the path of an electron from one place to another is to take into consideration every path that it could possibly take to get there. Every possibility must be taken into account, and their individual probabilities must be added together in order to mathematically explain what is actually observed. Electrons (and photons), it would seem, do not always move in perfectly straight lines, but instead wander about in strange, wavy, curving paths, emitting and absorbing photons and changing both their speed and direction. Only when all of these possible paths add together (and they are literally infinite in number) do we find that the path of least time is (usually) the path that is observed.

This method of treating all possibilities as reality, which Feynman called the "sum over histories" approach, is much more than just a clever trick. It seems to be the truth, as it has been thoroughly tested experimentally.

### FEYNMAN DIAGRAMS

One of Feynman's most memorable contributions to science, and particularly to QED, is the Feynman diagram. This diagram simplifies things by way of converting the normal four dimensions of space-time into a much simpler two dimensional graph—the two dimensions are merely space and time. These diagrams allow theorists to better visualize particular actions and interactions and then to calculate the probabilities of these events using some very intuitive mathematical tricks developed by Feynman (and refined by others). Though the mathematics of QED is undeniably difficult, these handy little diagrams do much to help the medicine go down a bit more easily.

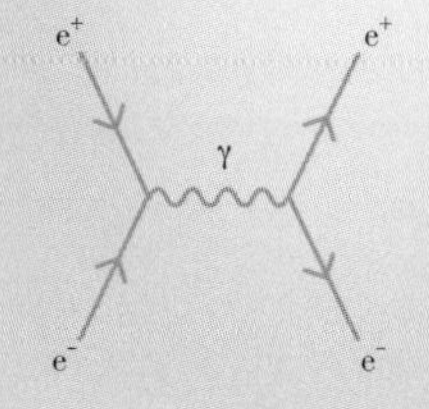


• Feynman simplified every physical event into just two dimensions—space and time. Space is the horizontal axis and time is vertical.

# PARTICLE ACCELERATION AND THE FUTURE OF PHYSICS

In late 2009 the largest particle accelerator ever built (indeed, the largest scientific instrument ever made), the Large Hadron Collider, went into operation on the Swiss-French border. This achievement capped off nearly a century of innovative devices which have led us to understand the world of particles on a far deeper level than we could have ever imagined.

### The Theory of Acceleration

Einstein's famous equation, $E = mc^2$ (see pp. 116–117), states that pure energy can theoretically be transformed into matter. Experimentation showed physicists that with enough energy, any particle could be created. The more energy, the more massive the particles. This is the theory behind all of particle physics.

Any physicist attempting to find a particle simply needs to produce the amount of energy necessary for the particle to be created. This is what particle accelerators do. They speed up particles and slam them together, producing energy out of which second-generation particles are created. These can be detected and analyzed using a number of devices, such as the cloud chamber (a pressurized chamber in which a particle's ionization trails can be photographed), the bubble chamber (in which particles leave trails of bubbles in a liquid), or the more modern spark chambers or proportional wire detectors which use computer-assisted detection.

### The Earliest Accelerator

In 1912 Austrian-American physicist Victor Hess first carried detectors with him in a hot air balloon and discovered cosmic rays. These rays of particles shower Earth from all corners of outer space at unbelievably high speeds, originating from sources such as solar flares, distant supernovas, or nearby stars.

Cosmic rays travel at speeds very close to the speed of light, smashing into the particles in our atmosphere and resulting in the formation of secondary particles which can be detected here on Earth. The limitation is that there is no way of controlling just when and where these rays will strike.

### Man-Made Accelerators

The first artificial particle accelerator, the cyclotron, was invented in 1929 by Ernest Lawrence at the University of California, Berkley. Particle accelerators such as this fire a beam of particles at high speed, which collide with other particles inside a detector.

Today, particle accelerators come in two standard varieties: linear and circular. Linear accelerators require a perfectly straight beam of particles, which is accelerated using finely-tuned electromagnets of increasing power. The particles are sent at high speeds into other particles near a detector, where the results are recorded and analyzed. The longest linear accelerator in the world today is the two-mile-long Stanford Linear Accelerator Center in Menlo Park, California.

Circular accelerators are perhaps even more impressive. The largest such accelerator is the new Large Hadron Collider (LHC) in Europe. The LHC is the most ambitious project (and expensive, at $2–5 billion) of its kind ever undertaken, and will hopefully be used to discover several new particles which have been theorized, such as a particularly elusive particle known as the Higgs boson (see p.161).

Circular accelerators confer the benefits of an infinitely long distance

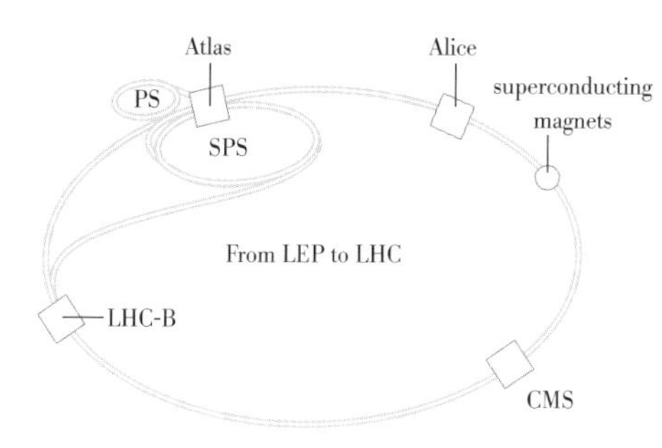


• A map of the Large Hadron Collider, currently operating on the Swiss/French border—the most powerful science experiment ever conducted.

through which the beam may be accelerated—particles may potentially travel through the same system forever, accelerating faster and faster (closer and closer to the speed of light) until there is no longer enough power to accelerate it any further.

Within these colossal machines are potential answers to many of the universe's mysteries, such as why particles have mass, how all the particles and forces are related to one another, and just what the early universe looked like. We may very well be standing at the doorstep of some of these answers right now!

## MEDICAL ACCELERATORS

Cyclotrons are still used today in some hospitals. Finely tuned particle beams from these machines have been frequently used to target and destroy cancerous tissue in patients. Accelerators are also in positron emission tomography (PET), which uses positrons produced in small accelerators to detect the emission of gamma rays inside the human body in order to reconstruct three dimensional images of a body's internal structures. In many ways, physics is invaluable in modern medicine.

# THE PARTICLE ZOO

In the biological sciences, there is an important field of research known as taxonomy (derived from the Greek *taxis*, meaning "the law of order"). Taxonomists attempt to classify plants and animals according to their various features, into such categories as kingdom, phylum, genus, and species. Only in the past half-century has a very similar field of study arisen in physics, which we might simply call atomic taxonomy—a push toward classification of the many subatomic particles.

**The Glory Days of Particle Physics**

The middle decades of the twentieth century were some of the most fruitful ever known in physics. They saw the creation of progressively more effective methods of particle acceleration and detection, along with the discovery of ever more complicated and unexpected physical laws.

Year by year, decade by decade, dozens upon dozens of new particles were being discovered, classified, studied, and pondered over. Sometimes a new theory would lead to the search for a new particle (as was the case with the positron in 1932 and the pion in 1947), while at other points a random discovery of a new particle would lead physicists on a search for a theoretical explanation.

Though particles may be classified in numerous ways—for example, by mass and charge—today it is most common to classify them all according to three distinct groups: leptons, mesons, and baryons.

**Leptons**

Leptons were originally considered to be the lightest of the atomic particles (the name comes from the Greek word *leptos*, meaning "light ones"). The name was applied to these particles before it was understood that some leptons are really not very light at all.

Today we believe that six total leptons exist—the electron, the muon, the tau, the electron neutrino, the muon neutrino, and the tau neutrino. Though of these only the electron and the electron neutrino is common in the universe, the others (such as the muon and the tau, which are essentially just really heavy electrons) have been found using particle accelerators or in high speed cosmic ray collisions in Earth's atmosphere.

The neutrinos are extremely light, neutrally charged particles which arise out of certain nuclear processes, such as radioactive decay. They are notably difficult to study experimentally, though we have managed to find clever means of finding them.

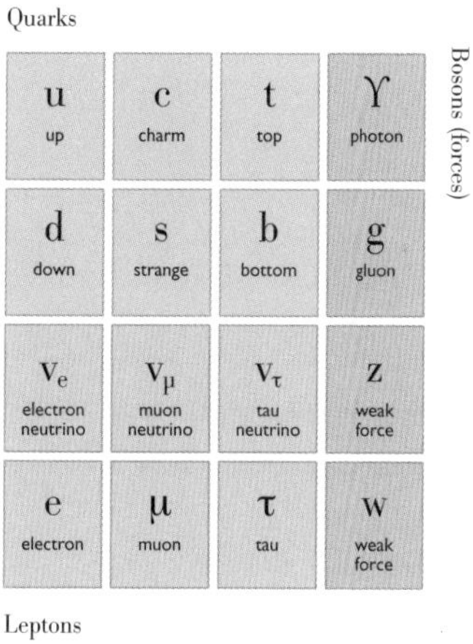

Quarks

| | | | | |
|---|---|---|---|---|
| u<br>up | c<br>charm | t<br>top | γ<br>photon | Bosons (forces) |
| d<br>down | s<br>strange | b<br>bottom | g<br>gluon | |
| $\nu_e$<br>electron neutrino | $\nu_\mu$<br>muon neutrino | $\nu_\tau$<br>tau neutrino | Z<br>weak force | |
| e<br>electron | μ<br>muon | τ<br>tau | W<br>weak force | |

Leptons

• The twelve major particles in modern physics, divided into "generational" columns, together with their respective force-carrying bosons.

## Mesons

Next step up from leptons are the mesons (with a name which is derived from the Greek word *mesos*, meaning "middle"). Mesons often act as particles which carry force between other particles. Among the mesons are the pion (which carries force between protons and neutrons, allowing the atomic nucleus to stick together), the kaon, the j/psi and many more (physicists have discovered well over 20 other mesons, though most of them possess lifetimes far too short to serve any real purpose).

## Baryons

The final particles are known as baryons (with a name meaning "to burden," or "weigh down" in Greek). This family consists of the familiar nucleons—the proton and the neutron—along with more than a hundred lesser-known and seemingly inconsequential particles (plus antiparticles) such as the Delta, Xi, Lambda, Sigma, and Omega, each of which come in a wide array of varieties (the Xi particle alone comes in at least twenty varieties).

Though the sheer number of baryons might seem imposing on the surface, we may take comfort in the fact that only a couple of them—the proton and the neutron—really serve any purpose in our everyday lives. The others exist, it seems, only to add beauty to our theories.

## Quarks

In the late 1960s American physicist Murray Gell-Mann simplified matters considerably by discovering that two of these "families" of particles—the mesons and the baryons—are actually composed of even smaller particles, called quarks, which are only six in number (just as there are six total leptons). They are: up quark, down quark, strange quark, charmed quark, top quark, and bottom quark.

All mesons are made up of two quarks—one regular quark and one antiquark (for antimatter versions of quarks are known to exist, just like every other known particle). All of the baryons are made up of three regular quarks. For example, a proton is made up of two up quarks and one down quark, while a neutron is made of two down quarks and one up quark.

In addition to the six leptons and six quarks, there are a few force-carrying particles (such as the photon), all of which constitute the entire particle zoo!

# THE STANDARD MODEL OF PARTICLE PHYSICS

The standard model is, at present, the culmination of all that particle physicists have come to know about the smallest things in the universe. Under this model we have succeeded in simplifying absolutely everything in the physical universe down to just two dozen particles and four total forces. Though there remain a few holes in this theory, the standard model stands today as the most successful, complete model of physics ever developed. But the future, as always, is uncertain.

## The Forces

The simplicity of the standard model comes mainly in the fact that every action and reaction in the entire universe can be reduced down to the work of just four forces, each of which are very distinct, both in strength and function:

**The Strong Nuclear Force** is, appropriately, the strongest of all forces, and is responsible for holding together the atomic nucleus. It is the force by which quarks are held together within protons and neutrons and the force by which the protons and neutrons themselves are bound to one another. Though the strength of the nuclear force is far beyond anything that we experience every day, it is only effective over a very short range (that is, about the size of the atomic nucleus, appropriately).

**The Electromagnetic Force** is readily present and visible everywhere you look. It provides us with power, of course, but it is also the very reason that matter exists in the first place. It is this force by which atoms bind together to form all compounds and substances. We all consist of little more than atoms held together electromagnetically. We ought to be grateful, therefore, that electromagnetism is as strong as it is.

**The Weak Nuclear Force** is perhaps the most peculiar of the forces. Its purpose is not to hold things together, but rather to break them apart. It is this force which causes atoms to be unstable, to decay, and to split apart. It is because of the weak nuclear force that nuclear weapons and energy are possible and, more importantly, it is the weak force which makes it possible for the Sun to shine and provide us with the necessary heat and light for survival. Like the strong force,

• The current atomic model, in which electrons orbit the nucleus. The nucleus is made up of protons and neutrons, which are themselves made of quarks.

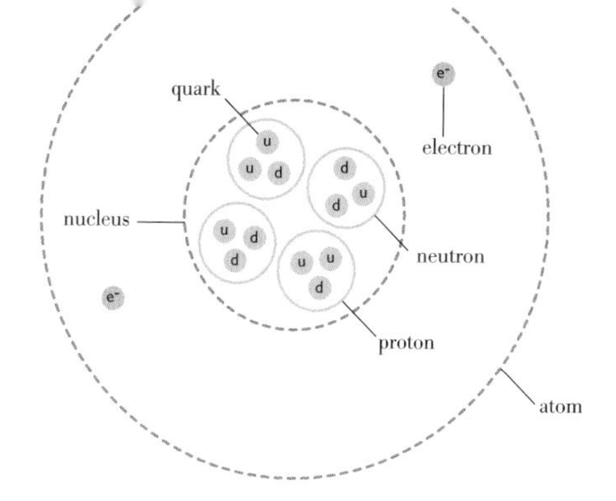


the weak force is only effective at very short distances within the atomic nucleus, and is just a bit weaker than electromagnetism.

**Gravity**, by far the weakest of all the forces (at $10^{39}$ times weaker than the strong nuclear force), is surely the force which we are most familiar with. Though gravity is the most well-known force, it is also perhaps the least understood. Gravity has proven monumentally difficult to factor into the standard model, and is the main reason this model is considered a work-in-progress.

## The Particles

In the standard model as it is presently formulated there are 17 total particles (29 once the antiparticles are accounted for). These particles can be divided into three groups:

**Leptons** – the six known leptons include the common electron and its identical-but-larger cousins, the muon and the tau. Each of these particles also possesses a neutrino counterpart—the electron neutrino, muon neutrino and tau neutrino. In the standard model each of these pairs belongs to a particular generation. First-generation leptons include the electron and electron neutrino; the second generation are the muon and muon neutrino; and the third generation are the tau and tau neutrino.

**Quarks** – Like the leptons, the six quarks are also divided into generations according to their differing masses. The first generation of quarks includes the two most common, the up quark and the down quark. The second generation consists of the strange quark and charmed quark, and the final generation consists of the top quark and the bottom quark. The branch of physics that deals with quarks is known as quantum chromodynamics.

**Gauge Bosons** – gauge bosons are the "force-carrying" particles of the subatomic world. Each force is associated with its own gauge bosons which can be exchanged between the other particles in order to keep them stuck together (or to pull them apart). These gauge bosons consist of the photon which carries the electromagnetic force, gluons which carry the strong nuclear force, and the W and Z particles which convey the weak nuclear force. It is believed that at least two (or more!) gauge bosons exist which have not yet been discovered, including the graviton which conveys the gravitational force and the Higgs boson, which, if discovered, is thought to be the particle responsible for giving mass to all of the other particles.

Exercise 15

# Escape Velocity

## THE PROBLEM:

One of the most important things that any rocket scientist must keep in mind when determining if a launch will be successful or not is the notion of escape velocity. Scientists must calculate the speed required to successfully escape Earth's gravitational forces and design the propulsion systems of their rockets accordingly. So what speed must a space shuttle attain to break free from the Earth's gravity?

## THE METHOD:

First we need to determine the escape velocity of Earth. To do this we use the standard formula that physicists have formulated to calculate exactly what the escape velocity is for a given body. That equation looks like this:

$$V_{esc} = \sqrt{\frac{2GM}{R}}$$

In this formula, G is the gravitational constant; M is the mass of the larger body (in this case, Earth); R is the distance between the two objects; and $V_{esc}$ is the velocity an object must attain to escape this gravitational force.

One thing to note about this equation is its similarity to Newton's famous law of gravitational attraction, except that here we are only concerned with one mass—that of the body exerting the strongest gravitational attraction (in the case of a rocket leaving Earth, this is would be Earth). We don't need to know the mass of the rocket itself because the escape velocity is the same for an object of any size. Where the mass becomes important once again is in determining just how much energy will be required to accelerate the rocket to such a speed.

Now, it should not be difficult to solve a very simple problem using just this formula.

## THE SOLUTION:

To find the escape velocity for Earth, we simply need to plug the appropriate numbers into the equation and solve it.

$$V_{esc} = \sqrt{\frac{2GM}{R}}$$

We can replace G with the best known value of the gravitational constant ($6.6743 \times 10^{-11}$ $m^3$ $kg^{-1}$ $s^{-2}$), M with the mass of Earth (which has been estimated at about $5.976 \times 10^{24}$ kilograms–no, you don't have to measure it yourself), and R with the distance between the center of Earth and our starting location (approximately $6.37 \times 10^6$ meters). This gives us:

$$V_{esc} = \sqrt{\frac{2[6.6743 \times 10^{-11} \times 5.976 \times 10^{24}]}{6.37 \times 10^6}}$$

Which can then be solved to arrive at a value for $V_{esc}$ of $1.12 \times 10^4$ meters per second, or 11.2 kilometers per second. Clearly this number represents a very high speed, and one which is certainly rather difficult to attain using conventional propulsion techniques from the surface of Earth. The scientists who develop outer space missions understand that to get a spacecraft to escape the force of Earth's gravity is a multi-step process. The first step is to get the spacecraft into orbit around Earth, where gravitational forces, air resistance, and friction are lower, making higher speeds far easier to obtain. It is from here that a spacecraft can utilize its propulsion systems once again in order to obtain the necessary speeds.

The escape velocities for the other planets in the Solar System vary according to their mass:

| | Mass ($\times 10^{21}$ kg) | Escape Velocity (km/s) |
|---|---|---|
| Sun | 1,989,100,000 | 618 |
| Mercury | 330.2 | 4.3 |
| Venus | 4,868.5 | 10.3 |
| Earth | 5,973.6 | 11.2 |
| Moon | 73.5 | 2.3 |
| Mars | 641.85 | 5 |
| Jupiter | 1,898,600 | 60 |
| Saturn | 568,460 | 36 |
| Uranus | 86,832 | 22 |
| Neptune | 102,430 | 24 |

# BLACK HOLES

Black holes are not particularly new to science—the possibility that black holes could exist was first considered in the eighteenth century. It wasn't until the latter half of the twentieth century that our understanding advanced significantly, when cosmologists began to look at the universe in detail through the lens of Einstein's general theory of relativity.

## What Is a Black Hole?

A black hole is not, in fact, a hole; rather, it is the opposite. Where a hole may be defined as the absence of substance, a black hole is a body of mass so very dense, with such a great gravitational attraction that nothing, not even light, may travel fast enough to escape its grasp. They are like astronomical whirlpools which suck in everything nearby.

Though it was the work of Einstein in 1915 that gave an insight into the details of black holes, credit for the first theoretical musing that such things might exist should go to the eighteenth-century geologist, John Michell.

Drawing on his understanding of escape velocity (as explained on the pp. 162–163), Michell considered that if an object had sufficient mass and thus an escape velocity greater than the speed at which light travels (about 300,000 km/s), then even light wouldn't be moving fast enough to escape from such an object. The object would be what is today known as a black hole.

Michell never really theorized that such an object truly did exist—at the time it was just an interesting idea.

## Black Holes Today

After the development of general relativity, the concept of black holes achieved an entirely new scientific foundation. Based on Einstein's accepted theory of gravity as warped space-time, scientists could explore how black holes might actually form from the collapse of large stars and what kind of effect this might have on nearby objects (which was thought to be the only way

"Black holes are regions of space where the gravity is so high that the fabric of space and time has curved back on itself, taking the exit doors with it."

***—Neil deGrasse Tyson, American astrophysicist***

they could be detected). In the 1920s, the German physicist Karl Schwarzschild explored the mathematics behind black holes and developed the Schwarzschild radius, a rather simple equation which showed how the diameter of a black hole changes by way of its mass.

The term "black hole" was only coined by Caltech physicist John Wheeler (to replace the previous term, "frozen star") in 1967, and it quickly caught on as others, such as Stephen Hawking (see pp. 168–169), took up the study of these peculiar objects. Hawking was the first physicist to recognize that black holes aren't entirely black with his discovery of what has become known as Hawking radiation.

While black holes have not actually been seen by way of this radiation at this point, their presence in the universe is very difficult to deny based on observations of their effects on other astronomical objects (large gravitational attractions with nothing at the center; and binary systems where the second object doesn't seem to exist, for example), and it is believed that super-massive black holes may act as the center of gravity for many (if not all) of the galaxies in the universe, including our own.

## Creating Black Holes

Today, particle accelerators have potentially become powerful enough that tiny black holes (micro black holes) may actually be produced on Earth! These tiny black holes are similar to those known as primordial black holes, which are said to have existed at the formation of the universe, and only exist for an exceedingly short period of time before disappearing—they evaporate by giving off Hawking radiation—thus giving physicists only an instant to explore their mysteries.

While there is some controversy regarding the wisdom of creating black holes on Earth, there is excitement amongst physicists as to where such explorations might lead in regard to deciphering some of the fundamental mysteries of the universe.

So today, as much as ever, black holes are of great scientific importance. Most likely, it will remain as such for a long time to come.

• Within Einstein's general theory of relativity, a black hole is a place where space-time becomes infinitely warped, creating a literal "hole" in space-time out of which nothing can escape.

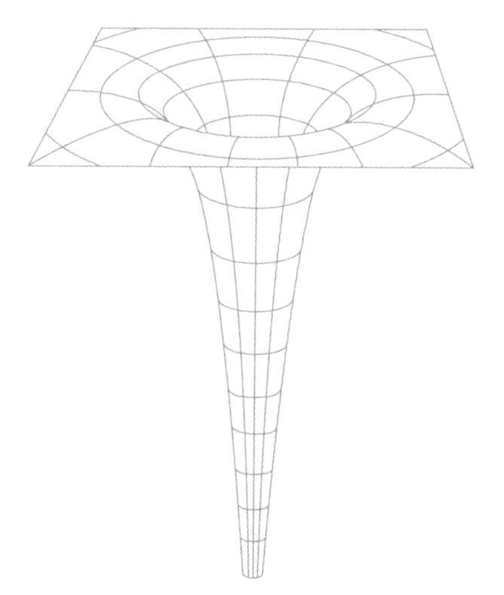

# The Schwarzschild Radius

## THE PROBLEM:

Karl Schwarzschild is fascinated by black holes. He knows that a black hole is in reality nothing more than an object which becomes so dense that nothing, not even light, may escape its gravitational attraction. He also knows that any object, regardless of its mass, would become a black hole if it were compressed to sufficient mass-density. His scientific curiosity leads him to ask some fascinating questions, such as: How much would the Sun have to be compressed in order to become a black hole? How about the Earth? How about a human being?

## THE METHOD:

Schwarzschild knows that there should be a relatively simple mathematical formula for determining the radius of any black hole's event horizon—this is the outermost edge of a black hole and the point of no return for light and any other object falling into it—accounting for only gravity and its effects on light. All that is needed to determine the size of this radius—the Schwarzschild radius—is the object's mass.

The resulting formula is simple:

$$r = \frac{2Gm}{c^2}$$

Where, $r$ is the radius of the black hole (the number we are trying to find), G is the gravitational constant (with a measured value of about 6,67 x $10^{-11}$ $Nm^2$ / $kg^2$), $m$ is any object's mass (in kilograms), and $c$ is the constant speed of light (as we've already seen, about 300,000 km/s).

## THE SOLUTION:

Schwarzschild now needs only to insert the known masses of the objects in question and solve the equation.

Fortunately, Schwarzschild did not have to measure the mass of the Sun himself—for he knew that it had been estimated to be somewhere around $1.98892 \times 10^{30}$ kilograms.

The mass of Earth is approximately $5.9742 \times 10^{24}$ kilograms.

The mass of the average human being is about 63 kilograms.

So, Schwarzschild solves the equation, first for the Sun:

$$r = \frac{2G(1.98892 \text{ x } 10^{30})}{c^2}$$

$$r = 2.5\text{km}$$

This means that the entire mass of the Sun would have to be crammed into a tiny ball just 2.5 kilometers across in order to have a great enough mass density to be considered a black hole. The size of such a black hole would be rather insignificant when compared to the largest black hole detected so far, which is about 18 billion times more massive than the Sun!

Plugging Earth's mass into the equation is even more extreme:

$$r = 0.0088\text{m}$$

This is just under nine millimeters. This means that the entire Earth would have to be reduced to a sphere just 9mm across in order to be considered a black hole!

What size black hole would the mass of an average human being translate to? The reader is encouraged to explore this question on their own.

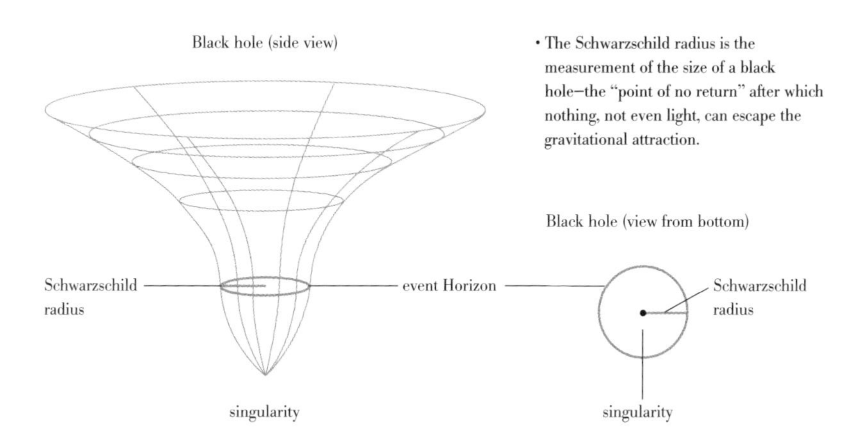


• The Schwarzschild radius is the measurement of the size of a black hole—the "point of no return" after which nothing, not even light, can escape the gravitational attraction.

Profile

# Stephen Hawking

**Most people know only a few things about Stephen Hawking—that he is one of today's most brilliant scientists and that he suffers from a rare and debilitating disease which has left him confined to a wheelchair and unable to speak but for the aid of a sophisticated computer. What is not commonly known, of course, is the depth of his contribution to the world of physics, both in the depth of his physical and mathematical knowledge and his desire to further the scientific dialogue. Hawking has been instrumental in inspiring popular interest in science.**

## Hawking's Life and Personal Tragedy

Stephen Hawking was born in Oxford on January 8, 1942. Though his parents permanently lived in London, the city was under constant bombardment from German bombers in a period known as the Blitz. Returning home to Highgate in north London after the war, Hawking began his schooling. His father, a medical researcher, urged him to pursue scientific studies, and after looking into both biology and medicine, Stephen eventually decided to pursue mathematics and physics.

In 1950 the Hawking family moved to St. Albans for the sake Stephen's father's career. Though Stephen wanted to study mathematics, his father urged him to study chemistry in order to attend Oxford University, where chemists had greater opportunities. In 1959 Hawking took the scholarship examinations at Oxford and was awarded a scholarship. He was awarded a First Class degree in 1962, with a specialization in physics. Hawking then attended Cambridge University, studying general relativity and cosmology. During this time, Hawking saw a doctor regarding some health peculiarities, including an increasing clumsiness he had begun to feel during his last year at Oxford.

In 1963 Stephen Hawking was diagnosed with Amyotrophic Lateral Sclerosis (ALS, or Lou Gehrig's Disease)—a degenerative condition that attacks the motor neurones, impairing the ability of the brain to communicate with the rest of the body—and doctors predicted that he would not live long enough to complete his

**Key work**

- **A Brief History of Time** *One of Hawking's most memorable achievements does not fall into the category of formal science at all, but instead is his publication of a book meant for general audiences– 1988's* A Brief History of Time. *This book, which covers such seemingly intimidating topics as general relativity, black holes, the big bang and predictive cosmology, is one of the best-selling books of all time, having sold over 9,000,000 copies since its release! The book is essential reading for anyone desiring a clear, concise explanation of modern cosmology from one of the greatest minds in that (or any other) field.*

• Dr. Stephen Hawking as he survives today: a brilliant mind trapped within, but not stifled by, a paralyzed body.

studies, so quickly was his health deteriorating. Hawking was married in 1965 to Jane Wilde, an event that he claims gave him the strength to continue fighting for his doctorate. Still, his health deteriorated and by the mid-'70s he had lost the ability to even feed himself. Ten years later, when he was confined to a wheelchair and almost entirely paralyzed, Hawking suffered from a bout of pneumonia and was given a tracheotomy, which left him unable to utter even the most basic sounds. Since then, Hawking's only communication has come by way of a voice synthesizer.

## Hawking's Scientific Achievements

Despite his personal hardships, Hawking persisted in studying cosmology, and as his condition finally began to stabilize, he pursued his degree with distinction, eventually receiving his PhD in 1966 and embarking on research throughout the next several decades which would contribute more to our understanding of gravity and the cosmos than any man since Albert Einstein.

Hawking's first major contribution came with his theory of singularities. He used Einstein's theory of general relativity (see pp. 118–121) to theorize that a point may technically exist in which both space and time become, essentially, infinite. This is the basic idea which governs our current understandings of both black holes and the point of space-time which existed just before the "Big Bang" brought our present universe into existence.

Hawking's theory of black holes led to important new facets of our understanding of these peculiar things. For example, according to quantum mechanics, black holes actually give off trace amounts of heat (which became known as Hawking radiation), which contradicted the previously held belief that nothing could ever escape from a black hole. He also predicted that black holes need not be immense, but may actually be tiny—even as small as subatomic particles.

Hawking was appointed Cambridge University's Lucasian Professor of Physics in 1979, holding this position until 2009. He remains alive and active in physics and in teaching science to this day—45 years and counting after his doctors told him he would not survive to complete his degree!

# A NEW COSMOLOGY

Cosmology is a branch of physics on a macroscopic scale. It is the physics of the universe as a whole; it is the shape, structure, behavior, past, present, and future of everything. It is only in the past few decades that astronomical observations have afforded physicists enough knowledge of the universe to begin solving some of these problems, though the search for answers has often just led us to even more questions.

## The Expanding Universe

One thing that seems relatively certain is that the universe is expanding. This does not mean merely that stars and galaxies are moving away from each other in space, but that the very fabric of space itself is in the process of expanding. Galaxies are moving away from each other as a result of the Big Bang—billions of years ago.

The expansion of the universe was discovered in the 1920s when Edwin Hubble noted that the color spectrum of distant galaxies seemed to be shifted slightly toward the red end of the spectrum. This was called a redshift—a telltale sign that galaxies were moving further apart.

This expanding view of the universe fits in well with the mathematics of general relativity. It gives us a more accurate description of the current state of the universe and provides insight into both the past and future.

• The three possible shapes of our universe: a "finite but boundless" sphere; a curved "saddle-shape"; or a nearly curveless shape.

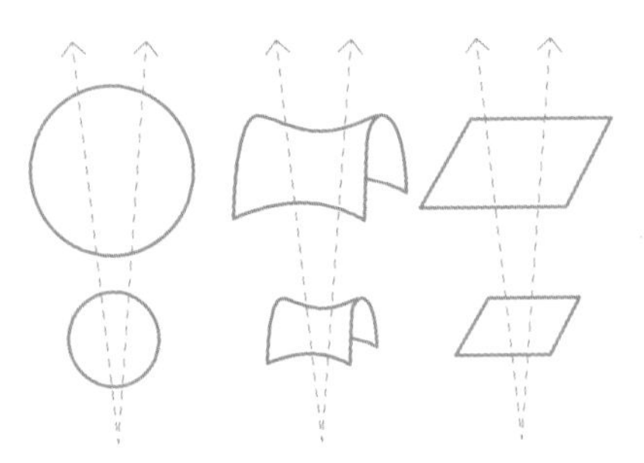

## The Past

To see the past, we need only to hit the rewind button, which is what physicists have been trying to do for more than half a century now. We know that if the universe is expanding now, then running time in reverse would lead to the universe contracting: stars, planets, and whole galaxies turned back into dense clouds of atoms and molecules, being pulled toward a single location by the all-pervasive force of gravity, their combined masses warping space-time and drawing greater amounts of mass into this location. Finally, after 10 or 20 billion years, everything comes together into a single, tiny area with near infinite density: a singularity. What may have gone on inside this infinitely dense state of matter is beyond the limits of our present knowledge, but out of it came all of matter, the laws of physics, and the dimensions of space-time themselves.

## The Future

In order to determine what will happen in the universe long after we are gone (we're talking many billions of years, most likely) we really only need concrete answers to two questions: How rapidly is the universe expanding at the current time? What is the average mass density of the universe?

The first question seems fairly simple and can be calculated to a certain extent using data from studying the retreat of distant stars and galaxies. The second question, however, is considerably more difficult. Obtaining even a rough estimate of how much mass exists in the universe for every square light year or so has proven harder than we could have imagined (see "Dark Matter and Dark Energy").

Finding answers to these questions would surely lead us to one of three conclusions:

1) The universe will continue to expand forever until all order in the universe is gone and all matter goes into a state of deep, permanent freeze.

2) The rate of expansion is slower and the universe will eventually be able to stop expanding and retain some sort of stability.

3) Eventually the expansion will stop and gravity will lead to contraction, pulling everything back together, leading to the universe collapsing in on itself to become a singularity once again (the big crunch).

Fortunately, there are enough unknowns to ensure the jobs of cosmologists for some time to come.

### DARK MATTER AND DARK ENERGY

A major stumbling block to measuring the density of the universe has been the discovery of mysterious and invisible substances which seem to permeate our universe, affectionately referred to as dark matter and dark energy.

Measurements taken of distant galaxies have revealed the presence of massive amounts of mass hidden somewhere in the universe which we can neither see nor detect. In fact, it is estimated that these dark substances constitute as much as 80 to 90% of all matter in the known universe! It is absolutely shocking that something so omnipresent could have escaped our notice for so long, but such seems to be the case. Today, many theories exist as to the nature of dark matter and dark energy, and it is hoped that continued research in particle accelerators might actually lead to some concrete answers, but for now it remains a mystery right out of science fiction.

# FINDING A THEORY OF EVERYTHING

It is, undeniably, the ultimate goal of physics to find a single, consistent theory which enables us to understand everything. This is what the dreams of physicists are made of and is something that many scientists, including Newton, Einstein, and many more, have attempted. It has been a long, hard couple of thousand years in the making, but could it be that we stand at the cusp of such a theory?

**The Problem of Gravity**

The standard model of physics seems like a pretty good start with its concise list of particles and forces, all of which together tell us just about everything we could want to know about why things are the way they are, but it is simply not yet complete. It is missing one extremely important addition: gravity.

Sure, gravity appears within the standard model's list of forces, but it is not explained within this model. Among the major problems with gravity's insertion into the standard model is the fact that it doesn't seem to fit. While the other three forces described in the standard model possess similarities which have led us to believe that there must be a deep underlying connection between them, we find no such convenient properties in gravity. Physicists have found good reason to believe that there is a point at which the nuclear forces and electromagnetism must converge and become one. The weak force and electromagnetism have indeed been combined into one force, known as the electroweak force. It has proven extremely difficult (if not impossible), however, to factor gravity into this model. This seems to be a particularly serious problem.

**The Origin of Mass**

Another question which, at the time of writing, has not been answered in full (though there seems to be some very real hope within the physics community that the answer might be arriving any day now) is the question of mass. Though mass has proven fairly simple to measure in many particles, we have not yet been able to determine exactly where this mass comes from. The best guess thus far is that it comes from the interaction of particles with an all-pervasive field in our universe known as the Higgs field (named for physicist Peter Higgs).

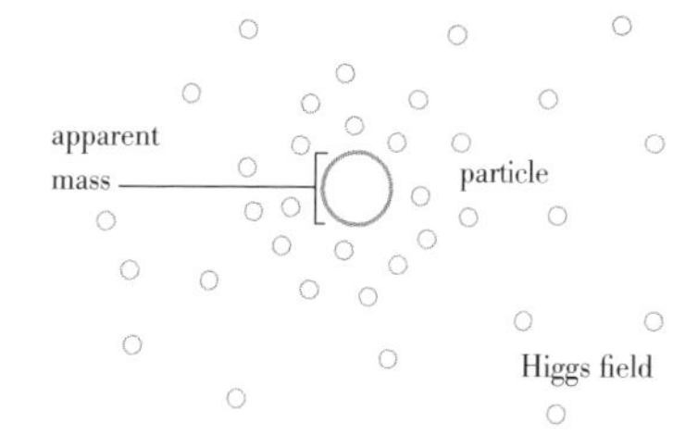


• In today's theories, the universe is dominated by a Higgs field. All particles travel through this field and interact with it, creating the "illusion" of mass.

If the Higgs theory is correct, then the Higgs field will naturally, like every other force, have an associated particle, known as the Higgs boson. It is this particle which experimenters hope to find in today's most modern particle accelerator, the Large Hadron Collider at CERN.

## Potential Theories of Everything

For decades physicists have been seeking out a theory of everything in numerous ways, some of which are exciting, while others seem somewhat more far-fetched. Some theories experience momentary fads and then fizzle out while others continually hover just beneath the radar as omnipresent possibilities.

String theory represents one of the most well-known of the potential theories of everything, though in actuality it encapsulates a multitude of overlapping ideas. String theory refers to any of several dozen potential models, each of which has just one thing in common: they believe that everything can be best understood as tiny, one-dimensional strings and their vibrations. That is string theory at its most simplistic; looping, knotting, dividing, vibrating strings which together form all space-time and matter in the universe, being the cause of all the great variation and order that we see in the universe around us. This all seems somewhat simple until it is realized that in many iterations of string theory this requires a universe of as many as ten dimensions in order to work! Not exactly intuitive. Other theories include quantum gravity, which requires gravity to possess its own messenger boson (the graviton) in an attempt to bring gravity into the standard model of physics by making it like the other forces. Another is supersymmetry, which declares that there is a perfect symmetry hidden deep within the laws of nature—a symmetry which existed immediately after the Big Bang but has since then been lost in an asymmetric universe. Under supersymmetric theories not only do all the known forces merge together into one, but the number of known particles doubles, for in supersymmetry every particle must possess a supersymmetric partner! It is the hope of many physicists that, like the Higgs boson, some of these supersymmetric particles might be within our grasp in the newest particle accelerators.

Perhaps none of these theories are correct, or perhaps only pieces of each are true. We simply don't know at this point, so we continue searching, for in physics, as in life, much of the fun lies in the journey rather than the destination.

# INDEX

# TERMS

Terms are explained where they are introduced within the text; however, a few are noted here for the sake of clarity.

**Antimatter** a theory stating that for every material particle in the standard model there exists a complementary particle with certain opposite properties, such as electric charge and spin.

**Atom** the fundamental unit of all tangible matter, consisting of a nucleus containing a mixture of protons and neutrons and surrounded by orbiting electrons.

**Black hole** any object so dense that its gravitational attraction allows neither matter nor electromagnetic radiation to escape.

**Electromagnetism** the field of physics concerning the behaviors of electric and magnetic waves.

**Energy** a measurement of an object or system's ability to do work. The motion, or change in motion, of an object is measured in terms of energy.

**Force** according to Newton's laws, that which causes a change in an object's motion. In the standard model, the four fundamental forces are: strong force, electromagnetic force, weak force, and gravitational force.

**Gravity** the fundamental force that is emanated from all matter and allows the accumulation of celestial bodies.

**Matter** a collection of the fundamental particles of physics. The opposite of matter is either pure electromagnetic radiation or a perfect vacuum.

**Particle Accelerator** a device that employs powerful magnets to accelerate fundamental particles and force them to collide, resulting in the production of secondary particles.

**Quantum Physics** a system developed in the twentieth century based on the notion that matter is best described using rules of probability and statistics rather than exact measurements.

**Radioactivity** the phenomenon in which an unstable atomic nucleus decays, emitting either alpha, beta, or gamma particles.

**Relativity** divided into Albert Einstein's "special" and "general" theories, the former is a four-dimensional model redefining the physical notions of both space and time and the latter a refined theory of gravitation.

**Scientific method** the logical systems by which scientific hypotheses and theories are developed.

**Standard model** the latest atomic model that incorporates the most tested physical theories and consists of four fundamental forces and 16 fundamental particles.

**Thermodynamics** a series of laws by which properties of heat and energy are defined.